AF760279

EDUCATION IN A COMPETITIVE AND GLOBALIZING WORLD

NATIONAL SCIENCE FOUNDATION

BACKGROUND, POLICIES, AND CONTRACTING PRACTICES

EDUCATION IN A COMPETITIVE AND GLOBALIZING WORLD

Additional books in this series can be found on Nova's website under the Series tab.

Additional e-books in this series can be found on Nova's website under the e-book tab.

EDUCATION IN A COMPETITIVE AND GLOBALIZING WORLD

NATIONAL SCIENCE FOUNDATION

BACKGROUND, POLICIES, AND CONTRACTING PRACTICES

LESLIE MURPHY
EDITOR

nova publishers
New York

Copyright © 2014 by Nova Science Publishers, Inc.

All rights reserved. No part of this book may be reproduced, stored in a retrieval system or transmitted in any form or by any means: electronic, electrostatic, magnetic, tape, mechanical photocopying, recording or otherwise without the written permission of the Publisher.

For permission to use material from this book please contact us:
Telephone 631-231-7269; Fax 631-231-8175
Web Site: http://www.novapublishers.com

NOTICE TO THE READER

The Publisher has taken reasonable care in the preparation of this book, but makes no expressed or implied warranty of any kind and assumes no responsibility for any errors or omissions. No liability is assumed for incidental or consequential damages in connection with or arising out of information contained in this book. The Publisher shall not be liable for any special, consequential, or exemplary damages resulting, in whole or in part, from the readers' use of, or reliance upon, this material. Any parts of this book based on government reports are so indicated and copyright is claimed for those parts to the extent applicable to compilations of such works.

Independent verification should be sought for any data, advice or recommendations contained in this book. In addition, no responsibility is assumed by the publisher for any injury and/or damage to persons or property arising from any methods, products, instructions, ideas or otherwise contained in this publication.

This publication is designed to provide accurate and authoritative information with regard to the subject matter covered herein. It is sold with the clear understanding that the Publisher is not engaged in rendering legal or any other professional services. If legal or any other expert assistance is required, the services of a competent person should be sought. FROM A DECLARATION OF PARTICIPANTS JOINTLY ADOPTED BY A COMMITTEE OF THE AMERICAN BAR ASSOCIATION AND A COMMITTEE OF PUBLISHERS.

Additional color graphics may be available in the e-book version of this book.

Library of Congress Cataloging-in-Publication Data

ISBN: 978-1-63321-972-4

Published by Nova Science Publishers, Inc. † New York

CONTENTS

PREFACE

This book includes information about the NSF for readers seeking an introduction to the foundation and its work. It is intended to provide background and institutional context for ongoing congressional consideration of NSF policy and fiscal issues. It also analyzes NSF funding trends and selected closely related STEM education policy issues in order to place conversations about NSF's budget in a broader fiscal and policy context.

Chapter 1 - The National Science Foundation (NSF) supports both basic research and education in the nonmedical sciences and engineering. Congress established the foundation as an independent federal agency in 1950 and directed it to "promote the progress of science; to advance the national health, prosperity, and welfare; to secure the national defense; and for other purposes." The NSF is a primary source of federal support for U.S. university research, especially in certain fields such as mathematics and computer science. It is also responsible for significant shares of the federal science, technology, engineering, and mathematics (STEM) education program portfolio and federal STEM student aid and support.

The NSF is an independent federal agency. Although governed by the budget and oversight processes, NSF's independent status has provided it with greater institutional autonomy than some other federal agencies. Some analysts assert that autonomy protects NSF's scientific mission. However, it may also be perceived as existing in tension with other public values, such as accountability. The tension between independence and accountability is an enduring policy theme for the NSF. It is reflected in debates over the foundation's authorization period and the role (if any) of Congress in grant-making and research prioritization.

NSF is the second-largest source of federal funding for basic research. Since FY2007, increases in the NSF budget have been driven by the doubling policy for physical sciences and engineering (PS&E) research. The PS&E doubling policy sought to double funding for targeted accounts at the National Institute of Standards and Technology, the Department of Energy's Office of Science, and NSF (total). Although this policy was authorized and reauthorized in the America COMPETES Act (P.L. 110-69) and America COMPETES Reauthorization Act of 2010 (P.L. 111- 358), and was pursued by both the George W. Bush and Obama Administrations, actual appropriations have not met authorized levels in most fiscal years. (Funding increased, just not to authorized levels.) PS&E doubling provisions expired in FY2013. President Obama did not seek funding for a PS&E doubling in FY2014 or FY2015. Bills to reauthorize NSF and other targeted accounts in the 113th Congress (specifically, H.R. 4186 and H.R. 4159) would authorize funding for some or all of the targeted accounts at growth rates that are below the paces set by the COMPETES acts.

In addition to its research responsibilities, NSF is the only federal agency whose primary mission includes education across all fields of science and engineering. (In this respect, NSF has a dual mission: research and education.) Funding for STEM education activities at NSF typically constitutes about a third of the total federal STEM education effort. Key questions for the 113th Congress focus on the Obama Administration's overall effort to reorganize the federal STEM education effort and the consequences of those changes for STEM education programs at NSF; the direction of the overarching federal STEM education strategy and NSF's role therein; as well as funding for STEM education at the foundation, as a percentage of total NSF appropriations.

NSF received $7.172 billion in funding (estimated) in FY2014. Typically, about 80% of the NSF budget supports the main research account, 12% supports the main education account, 3% to 5% supports facilities and construction, and the remainder supports administrative and related activities. Since FY2006, NSF appropriations have usually been included in annual Commerce, Justice, Science and Related Agencies appropriations acts. Major NSF authorizations expired in FY2013. At least two bills to reauthorize the foundation have been introduced in the 113th Congress (H.R. 4186 and H.R. 4159).

Chapter 2 - Federal policy makers have a long-standing interest in science, technology, engineering, and mathematics (STEM) education that dates to at least the 1st Congress. This interest is largely driven by concerns about the national science and engineering workforce, which is widely believed to play a central role in U.S. global economic competitiveness and national security.

The National Science Foundation (NSF) is a key component of the federal STEM education effort. Several inventories of the federal STEM education portfolio have highlighted NSF's important role—both in terms of funding and in the number and breadth of NSF programs. The NSF is also the only federal agency whose primary mission includes supporting education across all fields of science and engineering. As such, funding for STEM education at the NSF impacts not only the agency, but also the entire federal STEM education effort.

Congress reduced enacted funding levels (from the prior year) for NSF's main education account in both FY2011 and FY2012. Those year-over-year reductions followed several years of varying funding, as well as changes in the overall distribution of the foundation budget that reduced funding for the main education account as a percentage of the total NSF budget. For the most part, these changes appear to result from a combination of holding the main education account more or less constant while applying most of the foundation's FY2003-FY2012 budget growth to the main research account. However, in constant dollar terms, it appears at least some of the increase in funding for research activities during the observed period may have come at the expense of education activities.

It is not clear if these funding changes reflect evolving congressional and Administration policy priorities and an intentional prioritization of research over educational activities at the NSF, or if they reflect the cumulative impact of funding decisions made in response to specific conditions in specific fiscal years that happen to have had this effect. Further, the significance of these changes for NSF's STEM education and research missions—and for the overall federal STEM effort— depends, in part, on how they fit within the broader policy context. In particular, it depends (among other things) on how policy makers perceive and assess the policy rationale behind STEM education funding at the NSF; the character of NSF's STEM education activities; the foundation's role in the federal STEM education portfolio; and the impact of changes in NSF's education account on the foundation's other primary mission, research.

This report analyzes NSF funding trends and selected closely related STEM education policy issues in order to place conversations about NSF's budget in a broader fiscal and policy context. It concludes with an analysis of potential policy options.

Chapter 3 - The Experimental Program to Stimulate Competitive Research (EPSCoR) of the National Science Foundation (NSF) was authorized by

Congress in 1978, partly in response to concerns in Congress and the concerns of some in academia and the scientific community about the geographic distribution of federal research and development (R&D) funds. It was argued that there was a concentration of federal R&D funds in large and wealthy states and universities, and that the continuation of such funding patterns might ensure a dichotomy between the "haves" and "have-nots."

EPSCoR began in 1979 with five states and funding of approximately $1.0 million. Currently, EPSCoR operates in 29 jurisdictions, including 27 states and the Commonwealth of Puerto Rico and the U.S. Virgin Islands. To 2006, the NSF had invested approximately $920.0 million in EPSCoR programs and activities. When established, it operated solely in the NSF. EPSCoR was expanded in the mid-1980s and early 1990s; by 1998, seven other agencies had established EPSCoR or EPSCoR-like programs.

EPSCoR is a university-oriented program, with the goal of identifying, developing, and utilizing the academic science and technology resources in a state that will lead to increased R&D competitiveness. The program is a partnership between NSF and a state to improve the R&D competitiveness through the state's academic science and technology (S&T) infrastructure. Eventually, it is hoped that those states receiving limited federal support would improve their ability to compete successfully for federal and private sector funds through the regular grant system.

On November 18, 2011, President Barack Obama signed into law the Commerce, Justice, Science, and Related Appropriations Act, FY2012, P.L. 112-55. The law provides, among other things, funding for the NSF. The law provides a total of $7,033.1 million for the NSF in FY2012, $173.2 million above the FY2011 enacted level. Included in the total funding for NSF is $150.9 million for EPSCoR, approximately $5.5 million above the FY2011 level. The FY2012 appropriation for EPSCoR supports a portfolio of three complementary investment strategies— research infrastructure improvement ($110.0 million), co-funding ($39.4 million), and outreach and workshops ($1.5 million). It is anticipated that approximately 24.0% of the funding for EPSCoR in FY2012 will be used for new research awards. The remaining will be directed at providing support for grants made in previous years.

The Administration's FY2013 budget request for the NSF is $7,373.1 million, a 4.8% increase ($340.00 million) over the FY2012 estimate of $7,033.1 million. Included in the request total is $158.2 million for EPSCoR, an increase of 4.8% over the FY2012 estimate. Approximately 40.0% of the requested funding for EPSCoR in FY2013 will be directed toward new awards. The balance will support continuing awards made in prior years.

Chapter 4 - The NSF spends more than $400 million of its $7 billion annual budget acquiring goods and services in support of its mission to promote science and engineering. Much of this spending involves exploration activities in remote locations throughout the world, such as the Arctic and Antarctic. GAO examined the extent to which NSF uses key contracting practices in three phases of the acquisition process: (a) acquisition planning, (b) contract award, and (c) post-award contract monitoring. GAO selected and reviewed a nongeneralizable sample of 11 contracts or orders with at least $3 million in funding obligations for fiscal year 2011, which accounted for about 70 percent of NSF's total contract obligations for that year. Although all 11 contracts and orders received funding during fiscal year 2011, some were awarded more than 7 years ago. Some were awarded more recently. The authors reviewed each of the 11 contracts to determine the extent to which they reflected the use of key contracting practices based on the Federal Acquisition Regulation, the author's prior work, and NSF-OIG findings. GAO also reviewed NSF contracting policies and met with NSF contracting and program officials.

In: National Science Foundation
Editor: Leslie Murphy
ISBN: 978-1-63321-972-4
© 2014 Nova Science Publishers, Inc.

Chapter 1

THE NATIONAL SCIENCE FOUNDATION: BACKGROUND AND SELECTED POLICY ISSUES*

Heather B. Gonzalez

SUMMARY

The National Science Foundation (NSF) supports both basic research and education in the nonmedical sciences and engineering. Congress established the foundation as an independent federal agency in 1950 and directed it to "promote the progress of science; to advance the national health, prosperity, and welfare; to secure the national defense; and for other purposes." The NSF is a primary source of federal support for U.S. university research, especially in certain fields such as mathematics and computer science. It is also responsible for significant shares of the federal science, technology, engineering, and mathematics (STEM) education program portfolio and federal STEM student aid and support.

The NSF is an independent federal agency. Although governed by the budget and oversight processes, NSF's independent status has provided it with greater institutional autonomy than some other federal agencies. Some analysts assert that autonomy protects NSF's scientific mission. However, it may also be perceived as existing in tension with other public values, such as accountability. The tension between

* This is an edited, reformatted and augmented version of a Congressional Research Service publication, No. R43585, dated June 5, 2014.

independence and accountability is an enduring policy theme for the NSF. It is reflected in debates over the foundation's authorization period and the role (if any) of Congress in grant-making and research prioritization.

NSF is the second-largest source of federal funding for basic research. Since FY2007, increases in the NSF budget have been driven by the doubling policy for physical sciences and engineering (PS&E) research. The PS&E doubling policy sought to double funding for targeted accounts at the National Institute of Standards and Technology, the Department of Energy's Office of Science, and NSF (total). Although this policy was authorized and reauthorized in the America COMPETES Act (P.L. 110-69) and America COMPETES Reauthorization Act of 2010 (P.L. 111- 358), and was pursued by both the George W. Bush and Obama Administrations, actual appropriations have not met authorized levels in most fiscal years. (Funding increased, just not to authorized levels.) PS&E doubling provisions expired in FY2013. President Obama did not seek funding for a PS&E doubling in FY2014 or FY2015. Bills to reauthorize NSF and other targeted accounts in the 113th Congress (specifically, H.R. 4186 and H.R. 4159) would authorize funding for some or all of the targeted accounts at growth rates that are below the paces set by the COMPETES acts.

In addition to its research responsibilities, NSF is the only federal agency whose primary mission includes education across all fields of science and engineering. (In this respect, NSF has a dual mission: research and education.) Funding for STEM education activities at NSF typically constitutes about a third of the total federal STEM education effort. Key questions for the 113th Congress focus on the Obama Administration's overall effort to reorganize the federal STEM education effort and the consequences of those changes for STEM education programs at NSF; the direction of the overarching federal STEM education strategy and NSF's role therein; as well as funding for STEM education at the foundation, as a percentage of total NSF appropriations.

NSF received $7.172 billion in funding (estimated) in FY2014. Typically, about 80% of the NSF budget supports the main research account, 12% supports the main education account, 3% to 5% supports facilities and construction, and the remainder supports administrative and related activities. Since FY2006, NSF appropriations have usually been included in annual Commerce, Justice, Science and Related Agencies appropriations acts. Major NSF authorizations expired in FY2013. At least two bills to reauthorize the foundation have been introduced in the 113th Congress (H.R. 4186 and H.R. 4159).

INTRODUCTION

The National Science Foundation (NSF) supports both basic research and education in the nonmedical sciences and engineering. Congress established the foundation as an independent federal agency in 1950 and directed it to "promote the progress of science; to advance the national health, prosperity, and welfare; to secure the national defense; and for other purposes."[1] The NSF is a primary source of federal support for U.S. university research, especially in certain fields such as mathematics and computer science. It is also responsible for significant shares of the federal science, technology, engineering, and mathematics (STEM) education program portfolio and federal STEM student aid and support.

This report includes information about the NSF for readers seeking an introduction to the foundation and its work. It is intended to provide background and institutional context for ongoing congressional consideration of NSF policy and fiscal issues.

STRUCTURAL CHARACTERISTICS

Certain NSF structural characteristics set the foundation apart from other federal agencies and strongly influence its relationship with Congress. In particular, inventories of various federal agencies classify the NSF as an "independent agency." Two of the characteristics that contribute to this classification include NSF's position within the executive branch—it is freestanding, not within an executive department—and its leadership arrangement.[2] The NSF (unlike many other federal agencies) is governed by a 24-member board and a director, each of whom are appointed by the President to fixed, six-year terms.[3] Further, the foundation's organic act specifically establishes it as an "independent agency."[4] This independence, however, is not absolute. For example, the NSF's authorizing statute expressly references the President's authority to remove the director. Further, both Congress and the President retain the power to govern the NSF through the budget, appropriations, and oversight processes.

A Central Tension

In varying ways and to varying degrees, Congress has grappled with the tension between scientific independence and public accountability at the NSF since the foundation was established in 1950. (See section on "Legislative Origin.") This tension has remained a central policy theme for the NSF throughout its history. It is embedded in the very nature of the NSF as a federal entity and it underpins a wide variety of NSF policy debates—such as the debate about the length of foundation authorization periods (three years? five years? one?) and the debate about the NSF's grant-making process and merit-review criteria, which Congress has sought to influence on a number of occasions. Some policy makers assert that the foundation can best accomplish its scientific purposes if free from undue political influence, while others seek to ensure accountability in the expenditure of public funds. Each Congress has the opportunity to revisit this tension and to redefine the relationship between the NSF and Congress.

Policy makers express a variety of rationales for establishing independent agencies, including the belief that independence will facilitate better decision-making (particularly with respect to complex, ostensibly apolitical technical issues) or the desire to free agencies from the control and direction of the executive.[5] In the NSF's case, one NSF historian has observed, "Although the director was subject to removal by the President, his six-year statutory term, like that of the board members, *showed a desire to insulate the agency from politics* [emphasis added]."[6] Some analysts find trade-offs to agency independence though, noting that (in general), "autonomy can be a means of helping [agencies] accomplish democratic purposes ... however, [it] also shields them from direct accountability."[7] As a practical matter, legislators seeking to apply various federal assets toward specific national goals may find both benefits and barriers in the foundation's status as an independent agency.

LEADERSHIP AND STAFF

Consistent with the foundation's purposes, NSF leadership and staff include highly trained scientists and engineers from across the United States. More than half of NSF employees have earned at least a master's degree, over a quarter have a doctorate, and about 15% have completed post-doctorate education. In FY2012, about three-quarters of NSF staff held permanent appointments and about a quarter held non-permanent positions.

Leadership

The National Science Foundation is governed jointly by the NSF director and the 24- person National Science Board (NSB).[8] The director oversees the day-to-day activities of the foundation, including staff and management, program creation and administration, grant-making and merit review, planning, budget, and operations.[9] The board identifies issues critical to NSF's future, approves the foundation's strategic budget direction, approves annual budget submissions to the Office of Management and Budget, ensures balance between initiatives and core programs, and approves new major programs and awards.[10] The board also serves as an independent body of advisors to Congress and the President. NSF's director is an *ex officio* member of the board. NSB members typically come from industry and academia and represent a variety of STEM disciplines.[11] Historically, most NSF directors have come from the fields of physics or engineering.[12]

Appointment and Terms of Office

The President appoints the NSF director with the advice and consent of the Senate. The President also appoints the members of the National Science Board. (In 2012 Congress enacted legislation removing Senate confirmation requirements for the members of the NSB.)[13] Both the NSF director and members of the National Science Board serve six-year terms. NSB terms are staggered such that one-third of the board is appointed every two years.

Deputy Director

42 U.S.C. 1864a provides statutory authority for the NSF-wide deputy director and provides the deputy director with the power to act as NSF director in the event of a vacancy, disability, or absence. The deputy director also performs other duties as determined by the director.[14] Since the mid-1990s the deputy director has served as NSF's Chief Operating Officer. The President

appoints the NSF deputy director with the advice and consent of the Senate. The position includes no statutorily prescribed term of office.

Assistant Directors

The leaders of NSF's directorates carry the title "assistant director." The assistant director position is not currently statutorily authorized, but it has been in the past. In FY2014, there were eight assistant directors. Assistant director duties vary by directorate and in some cases have changed over the years. In general, assistant directors lead directorate programs and initiatives and are responsible for planning and implementing programs, priorities, and policies. Assistant directors are often non-permanent staff. In previous years, this position required presidential appointment and Senate confirmation.

Division Directors

Division directors are responsible for long-range planning and budgetary stewardship within their research areas. They also oversee the grant-making process and, in many cases, make the final programmatic decision to approve (or decline) awards to NSF grant-seekers.

Program Directors

Program directors are subject matter experts. They conduct the scientific, technical, and programmatic review and evaluation of proposals, including peer reviewer recruitment and management of the proposal review process. They manage program budgets and provide award oversight. Program directors make funding recommendations to division directors.

Rotators

The NSF workforce is made up of permanent, temporary, and "rotating" staff. Unlike permanent and temporary staff, most rotating staff are hired under the authority of the Intergovernmental Personnel Act of 1970 (IPA, P.L. 91-648) and are not considered federal employees. IPA rotators typically come from institutions of higher education, but they may come from other organizations as well (e.g., state and local government, Indian tribal government, nonprofit entities). IPA rotators retain ties to their home institutions—including pay and benefits— and may serve the NSF for no more than four years. Unlike its practices for permanent and temporary staff, NSF uses program funds to provide salary reimbursement, living expenses, and travel expenses for IPA rotators. Overall, IPA rotators comprised 12% of NSF's total workforce in August 2012. Further, of the 512 program directors at NSF in FY2012, 262 (51%) were permanent employees, 172 (34%) were IPA rotators, 39 (8%) were temporary employees, and 39 (8%) were visiting scholars.

NSF Rotators: Pros and Cons

Policy analysts debate NSF's use of rotators.[15] NSF asserts that rotators bring fresh, cutting-edge insight to foundation programs and that rotators increase knowledge transfer between the research community and the foundation. A 2004 National Academy of Public Administration panel largely substantiated these claims.[16] But NSF's Inspector General notes that IPA rotators are more expensive than federal employees and contends that NSF could enhance oversight and management of its rotator programs.[17]

Mission Activities

NSF's dual mission is to support basic research[18] and education in the non-medical sciences and engineering. NSF is the second-largest source of federal funding for basic research, and a top three source of federal funding for research in the fields of environmental sciences, life sciences, mathematics and computer sciences, physical sciences, social sciences, and other sciences.[19] Funding for STEM education activities at NSF typically constitutes about a third of the total federal STEM education effort.

The foundation currently divides its mission activities among seven directorates, which are mainly organized by academic discipline.[20] Until FY2013, the largest directorate (measured by budget authority) was Mathematical and Physical Sciences (MPS, $1.312 billion in FY2011 actual) and the smallest was Social, Behavioral, and Economic Sciences (SBE, $247.3 billion in FY2011 actual). NSF adjusted its organization chart in FY2013, placing some previously freestanding offices into existing directorates. As a result of these changes, the Directorate for Geosciences became the largest directorate at NSF. SBE remained the smallest. (See **Figure 1**.)

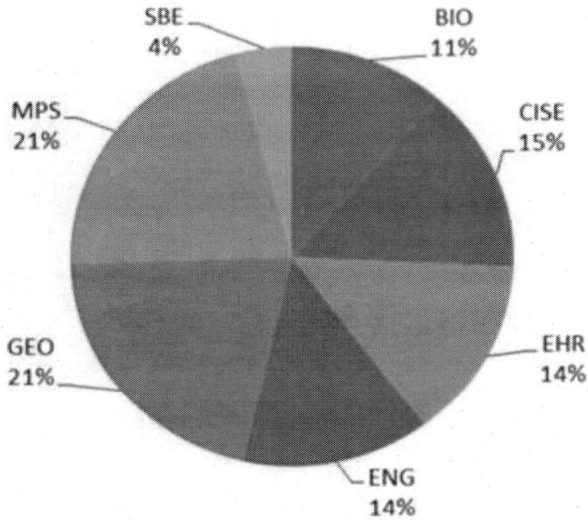

Source: Congressional Research Service, based on FY2013 actual funding levels reported in NSF's *FY2015 Budget Request to Congress*, available at http://www.nsf.gov/about/budget/fy2015/toc.jsp.

Notes: The term "SBE" refers to the Directorate for Social, Behavioral, and Economic Sciences; "BIO" refers to the Directorate for Biological Sciences; "CISE" refers to the Directorate for Computer and Information Science and Engineering; "EHR" refers to the Directorate for Education and Human Resources; "ENG" refers to the Directorate for Engineering; "GEO" refers to the Directorate for Geosciences; and "MPS" refers to the Directorate for Mathematical and Physical Sciences.

Figure 1. Distribution of Funding for NSF Mission Activities; FY2013 Actual, by Directorate.

NSF directorates are divided into divisions—with between four and six divisions per directorate, typically—that manage programs. The main appropriations account for all but one directorate is NSF's Research and Related Activities (RRA) account. The Directorate for Education and Human Resources is the exception to this rule; its main source of appropriations is the Education and Human Resources (EHR) account. Many NSF programs or projects are co-funded (i.e., they receive funding from two or more foundation accounts) or involve coordination and cooperation between programs and disciplines.

Directorate for Biological Sciences (BIO)

BIO's mission "is to enable discoveries for understanding life. BIO-supported research advances the frontiers of biological knowledge, increases our understanding of complex systems, and provides a theoretical basis for original research in many other scientific disciplines."[21] BIO divisions include Molecular and Cellular Biosciences; Integrative Organismal Systems; Environmental Biology, Biological Infrastructure; and Emerging Frontiers. In FY2013, BIO received $679 million in funding (actual).

Directorate for Computer and Information Science and Engineering (CISE)

CISE "supports investigator-initiated research in all areas of computer and information science and engineering, fosters broad interdisciplinary collaboration, helps develop and maintain cutting-edge national computing and information infrastructure for research and education, and contributes to the development of a computer and information technology workforce with skills essential for success in the increasingly competitive global market."[22] CISE divisions include Advanced Cyberinfrastructure, Computing and Communication Foundations, Computer and Network Systems, Information and Intelligent Systems, and Information Technology Research. FY2013 actual funding for CISE was $858 million.

Directorate for Education and Human Resources (EHR)

EHR seeks to "achieve excellence in U.S. science, technology, engineering and mathematics (STEM) education at all levels and in all settings (both formal and informal) in order to support the development of a diverse and well-prepared workforce of scientists, technicians, engineers, mathematicians and educators and a well-informed citizenry that have access to the ideas and tools of science and engineering."[23] EHR has four divisions: Research on Learning in Formal and Informal Settings, Graduate Education, Human Resource Development, and Undergraduate Education. FY2013 actual funding for EHR was $835 million.

Directorate for Engineering (ENG)

ENG "investments in engineering research and education aim to build and strengthen a national capacity for innovation that can lead over time to the creation of new shared wealth and a better quality of life."[24] ENG divisions include Chemical, Bioengineering, Environmental, and Transport Systems; Civil, Mechanical, and Manufacturing Innovation; Electrical, Communications, and Cyber Systems; Engineering Education and Centers; Industrial Innovation and Partnerships; and Emerging Frontiers in Research and Innovation. FY2013 actual funding for ENG was $820 million.

Directorate for Geosciences (GEO)

GEO's mission is "to support research in the atmospheric, earth, and ocean sciences."[25] GEO divisions include Atmospheric and Geospace Sciences, Earth Sciences, Integrative and Collaborative Education and Research, Ocean Sciences, and Polar Programs. FY2013 actual funding for GEO was $1.250 billion.

Directorate for Mathematical and Physical Sciences (MPS)

MPS's mission is to "harness the collective efforts of the mathematical and physical sciences communities to address the most compelling scientific questions, educate the future advanced high-tech workforce, and promote discoveries to meet the needs of the Nation."[26] MPS divisions include Astronomical Sciences, Chemistry, Materials Research, Mathematical Sciences, and Physics. FY2013 actual funding for MPS was $1.274 billion.

Directorate for Social, Behavioral, and Economic Sciences (SBE)

SBE's mission is "to promote the understanding of people and their lives by supporting research that reveals basic facets of human behavior; to encourage research that addresses important societal questions and problems; to work with other scientific disciplines to ensure that basic research and solutions to problems build upon the best multidisciplinary science; and to provide mission-critical statistical information about science and engineering

(S&E) in the U.S. and the world through the National Center for Science and Engineering Statistics (NCSES)."[27] In addition to the NCSES, SBE divisions include Social and Economic Sciences, as well as Behavioral and Cognitive Sciences. FY2013 actual funding for SBE was $243 million.

Funding for Social Science?

Congress has grappled with the question of funding for the social sciences since NSF was established. The debate generally centers on the question of public benefit. Those who seek to reduce funding for research in the social, behavioral, and economic (SBE) sciences typically assert that federal dollars should focus on fields perceived to be more closely associated with national security, health, or economic interests, such as the physical and life sciences. Some opponents also question certain NSF SBE grants—such as the "study of human-set forest fires 2,000 years ago in New Zealand"—which may be perceived as frivolous or wasteful.[28] Supporters, on the other hand, typically dispute the notion that SBE research does not serve the national interest, citing research that is perceived to provide broad public benefit, such as research to improve disaster response, facilitate kidney matching between donors and patients, and research that informs innovation.[29] Supporters maintain that SBE grants can yield critical innovations (though they may appear frivolous) or assert that NSF's peer-review process assures that only meritorious proposals are funded.

The floor debate over H.Amdt. 734—which amended H.R. 4660 (Commerce, Justice, Science and Related Agencies Appropriations Act, FY2015) to reduce funding for NSF's SBE directorate by $15.4 million and to redirect the funds to physical sciences and engineering—offers an example of this policy debate in the 113th Congress.[30] Similar debates have occurred in the context of deliberations over H.R. 4186 (Frontiers in Innovation, Research, Science, and Technology Act of 2014), which (among other things) seeks to prioritize funding for physical sciences and engineering research at the NSF. H.R. 4186 was reported favorably from the House Committee on Science, Space, and Technology on May 28, 2014.

Legislators have also focused more narrowly on the specific question of funding for NSF's Political Science program, which is part of SBE. Some legislators assert that political science research is extraneous to NSF's central mission and a waste of federal dollars.[31]

Legislators who hold this view introduced several provisions limiting funding for political science at NSF in the 112th and 113th Congresses. Most of these provisions have not been enacted. However, one provision limiting funding for NSF's Political Science program in FY2013 (Section 543 of P.L. 113-6, Consolidated and Further Continuing Appropriations Act, 2013) became law.[32]

Some of those who objected to Section 543 asserted that it threatened NSF's independence and therefore the integrity of the research the foundation supports.[33] They also contended that Section 543 put the decision about what research to fund in the hands of Congress rather than in the hands of the scientists who lead and manage NSF (and who are, they argue, better able to judge the intellectual merit of various research proposals). Similar provisions limiting funding for political science were not enacted in FY2014. Average annual funding for NSF's Political Science program is typically in the $9 million to $10 million range.

Grant-Making

Because most NSF funding is distributed to researchers and institutions outside of the foundation, grant-making is (arguably) the heart of what NSF does.[34] Grants can be either standard (i.e., full funding up-front) or continuing (i.e., incremental funding on a multi-year basis). NSF receives approximately 50,000 grant proposals annually.[35] Of these, between 20% and 25% typically receive funding. About 36,500 scientists and engineers participated in the merit review process as panelists and proposal reviewers in FY2013.

The vast majority of NSF grants are awarded through a competitive, merit-based assessment process.[36] The peer review stage of this process—in which external "peer" reviewers with subject matter expertise assess the merits of each grant proposal—is both widely lauded and closely watched by policy analysts.

Although peer review is perhaps the most well-known stage of NSF's grant-making process, peer review does not encompass the whole of the assessment process. Rather, the typical grant-making process for most NSF awards follows three phases.

- Phase 1: opportunity announced, proposals submitted, proposals received.[37]

- Phase 2: reviewers selected, peer review, program officer recommendation, division director review.[38]
- Phase 3: business review, award finalized.[39]

Put differently, most NSF proposals must survive at least five kinds of scrutiny. First, the initial assessment is for application completeness and conformance with NSF requirements. Second, if a proposal survives the initial assessment, then it is sent to three or more external subject matter experts for peer review.[40] Peer reviewers evaluate the proposal according to two broad criteria: intellectual merit and broader impacts.[41] According to the NSF,

> "*Intellectual Merit:* The Intellectual Merit criterion encompasses the potential to advance knowledge; and
>
> *Broader Impacts:* The Broader Impacts criterion encompasses the potential to benefit society and contribute to the achievement of specific, desired societal outcomes."[42]

Peer reviewers provide information about the merit of the proposal to the program director, who (third) considers the proposal in the context of the broader program portfolio and direction. Program directors are not bound by the recommendations of peer reviewers. Rather, the program director

> reviews the proposal and analyzes the input received from the external reviewers. In addition to the external reviews, Program [Directors] consider several factors in developing a portfolio of funded projects. For example, these factors might include different approaches to significant research and education questions; potential (with perhaps high risk) for transformational advances in a field; capacity building in a new and promising research area; or achievement of special program objectives. In addition, decisions on a given proposal are made considering both other current proposals and previously funded projects.[43]

Fourth, after the portfolio assessment, the program director submits his or her award recommendation to the division director, who examines the recommendations and typically makes the final programmatic decision to fund (or not).

Fifth, if the proposal survives programmatic review (including initial, peer, program, and division), it is sent to the Office of Budget, Finance, and Award Management (BFA). Analysts within the BFA conduct an assessment of the business, financial, and policy implications, and, if called for, issue the grant.

In addition, larger or "sensitive" awards may require further layers of review beyond those already described, including review by NSF senior management or the National Science Board.[44] This rule applies to all Major Research Equipment and Facilities Construction (MREFC) projects. NSB also establishes average annual award amounts (for each directorate or office) that trigger NSB review and approval requirements. Trigger amounts differ by directorate and ranged from $6.88 million (BIO) to $12.66 million (GEO) in FY2014.[45]

Besides grants, NSF also awards funding though other mechanisms, such as cooperative agreements, contracts, and competitions. In FY2013, NSF distributed 73% of its funding via grants, 22% by cooperative agreement, and 5% by contract.[46]

The Geography of Grants

University research is widely believed to contribute to state and regional economic development. As such, many policy makers seek to increase research funding for their local colleges and universities. However, of the over 900 institutions reporting at least $150,000 in R&D in FY2012, the top 100 accounted for 79% of total academic science and engineering R&D. This trend has held constant for at least two decades. In response, some policy makers have sought mechanisms to reduce concentration in the geographic distribution of federal research grants, which make up a large portion of academic R&D.[47]

To this end, Congress established the Experimental Program to Stimulate Competitive Research (EPSCoR) at NSF in 1978.[48] EPSCoR is NSF's only state-based program. It is designed to strengthen the research capacity of institutions located within "EPSCoR Jurisdictions"—that is, those states that have historically received limited federal R&D funding—so that they are able to compete more successfully for federal R&D funding. Since the NSF EPSCoR program was established at least five other federal agencies have launched EPSCoR programs. The 2014 EPSCoR Jurisdiction map from the EPSCoR Interagency Coordinating Committee, which NSF staffs and chairs, includes between 23 and 31 eligible U.S. states and territories (depending on the agency).[49]

At the beginning of the EPSCoR program, some questioned the length of time required for a state to improve its research infrastructure.[50]

A five-year limit was proposed, but that proved to be "... unrealistic, both substantively and politically."[51] Questions remain concerning the length of time states should receive EPSCoR support. Some analysts assert that some states and their institutions should assume more responsibility for building their research infrastructure and become less dependent on EPSCoR funds. They argue that some researchers and states have become comfortable with EPSCoR funding and are not being aggressive in graduating from the program. In three decades no state has graduated from NSF's EPSCoR program. The issue of graduation has generated considerable congressional interest. In 2013, The National Academies published a study of federal EPSCoR programs, including findings and recommendations related to graduation and other program aspects.[52]

SCIENTIFIC FACILITIES, INSTRUMENTS, AND EQUIPMENT

NSF does not typically directly operate laboratories or scientific facilities. However, the foundation provides operations and maintenance support to a wide array of scientific facilities, instruments, and equipment. NSF funding supports the National Center for Atmospheric Research, polar facilities and logistics, and the academic research fleet. More information about NSF support for scientific facilities, instruments, and equipment is typically found in the "Facilities" sections of NSF's annual budget requests to Congress.

NSF's Major Research Equipment and Facilities Construction (MREFC) account also provides funding for (typically) between four and six research facilities and equipment construction projects. These projects include international activities, such as the Atacama Large Millimeter Array (a large radio telescope located in northern Chile that was completed in 2013);[53] as well as U.S. projects, including certain ground-based astronomical telescopes and ecological and ocean observatory networks (which connect geographically distributed scientific facilities and instruments). MREFC funds typically support projects only during the construction phase. Project planning and design, as well as post-construction operations and maintenance, comes from Research and Related Activities (RRA).

MAJOR CONSTITUENCIES

Approximately three-quarters of NSF funds are awarded annually to colleges, universities, and academic consortia. The remainder of NSF's budget usually goes to private industry (about 13%), Federally Funded Research and Development Centers (around 3%), and other sources (about 3%). NSF typically provides about 50% (or more) of federal funding for academic basic research in computer science, biology, environmental sciences, mathematics, and social sciences. Further, about a third of all known federal funding for STEM education comes from NSF in a typical budget year. The foundation is a primary source of support for graduate student fellowships in the non-biomedical sciences and engineering.

In FY2013, NSF issued 10,844 new awards to 1,922 colleges, universities, and other institutions in 50 states, the District of Columbia, and three U.S. territories. The foundation supported an estimated 299,000 individuals, including researchers, postdoctoral associates, and other professionals; undergraduate and graduate students; and elementary and secondary school teachers and students. At least 212 Nobel laureates have received NSF support at some point in their careers. NSF support for informal science education and scientific literacy reaches many Americans—in museums, libraries, afterschool programs, and through the media—every year.

Broadening Participation

The demographic profile of the U.S. student-age population is changing. The youth population is more racially and ethnically diverse than previous generations of Americans. At the same time, women have attained majority status on many college campuses. Yet, these groups may be underrepresented in certain STEM fields. Some analysts assert that underrepresented groups are an underutilized resource, which could be tapped to help meet perceived demand for STEM-competencies in the U.S. workforce.[54]

General agreement about the problems posed by racial, ethnic, and gender disparities in STEM education and employment has not translated into widespread agreement on either the causes of underrepresentation or policy solutions. Further, causes and solutions may be different for different population subsets.

Broadening participation issues include faculty diversity, the potential for bias in grant-making, and "family friendly" work environments for scientists and engineers, as well as teacher quality in schools that serve minority students, parental involvement and support for STEM-interested youth, and access to STEM-related education opportunities and support programs. Broadening participation is not limited to race/ethnicity and gender, either. Studies have shown STEM achievement gaps by income and level of urbanization (e.g., rural, suburban, urban) as well.

NSF operates several dozen programs that seek to broaden participation. The Science and Engineering Equal Opportunities Act (P.L. 96-516), which was incorporated into the 1980 NSF reauthorization, (1) declared that it is U.S. policy to encourage underrepresented populations to participate in STEM, and (2) authorized NSF to establish programs for this purpose. As of FY2013, funding for NSF's broadening participation programs was $755 million (actual). Some of the most widely tracked NSF broadening participation programs provide funding to minority-serving institutions of higher education.

SELECTED AUTHORIZATIONS

The following sections describe NSF's legislative origins and the foundation's two most recent, enacted reauthorizations as part of the America COMPETES Act (P.L. 110-69) and the America COMPETES Reauthorization Act of 2010 (P.L. 111-358). The 113th Congress has begun the process of reauthorizing certain COMPETES provisions, including NSF provisions. (See section titled, "Reauthorization Activity in the 113th Congress".) **Table A-1** includes a list of selected NSF authorizations dating to the 1950s.

Legislative Origin

Many contemporary policy conversations about the NSF mirror the debate over the foundation's establishment. For example, the 113th Congress has debated the question of funding for social science at the NSF. This issue was also debated during the establishment of the foundation. Retelling the historical conversation, therefore, contextualizes today's deliberations and provides legislators with additional insight into the enduring nature of some of these conflicts. It also provides insight into how previous generations of policy makers resolved similar questions.

Historical accounts of the NSF frequently peg the foundation's genesis to a dialogue between two men: Senator Harley M. Kilgore and Vannevar Bush.[55] Senator Kilgore chaired the Senate Subcommittee on War Mobilization during and immediately after World War II. Bush was director of the Office of Scientific Research and Development (ORSD) as well as a science advisor to President Franklin Delano Roosevelt.[56] Between 1942 and 1945, Senator Kilgore's subcommittee held a series of hearings on government support for scientific research. That effort resulted in the July 23, 1945, introduction of S. 1297 (National Science Foundation Act of 1945), which would have established a National Science Foundation. Bush, on the other hand, authored an historic July 1945 report on post-war U.S. scientific research, *Science: An Endless Frontier*, which called for the creation of a National Research Foundation. On July 19, 1945, Senator Warren Magnuson introduced a bill, S. 1285 (National Research Foundation Act of 1945)—which was drafted in consultation with Bush and hewed closely to the proposal outlined in *Science*—to establish a National Research Foundation.[57]

Although Senator Kilgore and Senator Magnuson agreed on the goal of establishing a federal agency for the support of scientific research and their bills shared certain similarities, they promoted different approaches.[58] There was agreement, for example, that the foundation should provide scholarships, that it should support basic research, that it should have both a board and a director, and that it should be independent from other executive branch agencies.[59] Differences focused on five broad themes that would be very familiar to an NSF observer today. These include

- ownership of patents resulting from government research,
- inclusion of the social sciences,
- geographic distribution of funding,
- the extent to which the foundation should support applied research, and
- political and administrative control of the foundation.[60]

As drafted in August of 1945, S. 1297 and S. 1285 would have resolved these policy issues differently. Senator Kilgore's bill (S. 1297) envisioned a scientific foundation that was administered by a publically appointed director and advised by a board, that distributed funding and research findings broadly, and that defined the term "research and development" to include both theoretical exploration as well as the extension of investigation

> into practical application, including the preparation of plans, specifications, and standards for various goods and services, the undertaking of related economic and industrial studies, the experimental production and testing of models, and the building and operation of pilot plants.[61]

Senator Magnuson's bill (S. 1285), on the other hand, would have created a research foundation led by a publically appointed board that would select, direct, and supervise a director. The powers and duties of the foundation as described in S. 1285 include developing national science policies and support of basic research in the fields of mathematical, physical, and biological sciences. The bill does not include provisions for the broad distribution of funding, though it does authorize the publication and dissemination of research findings.

The differences between these approaches were not resolved in the 79th Congress. However, after two more years of debate Congress presented a bill to establish a National Science Foundation to President Harry S. Truman on July 25, 1947 (S. 526, National Science Foundation Act of 1947). Truman vetoed the bill. In his veto message Truman expressed two concerns. First, Truman asserted that S. 526 violated his appointment powers and raised questions about accountability because it did not provide for a presidentially appointed director. (S. 526 gave authority to appoint a director to the foundation.) Second, the President expressed conflict-of-interest concerns. As defined in S. 526, the foundation included 24 eminent scientists appointed by the President with the advice and consent of the Senate. These 24 scientists would determine who would receive foundation grants, which Truman perceived as a conflict of interest that "would inevitably give rise to suspicions of favoritism."[62]

In April 1950, Congress sent the President a new bill, S. 247 (National Science Foundation Act of 1950).[63] President Truman signed S. 247, which became P.L. 81-507 (hereinafter referred to as NSF's "organic act") on May 10, 1950.[64] NSF's organic act provided for an independent federal agency administered by a presidentially appointed board *and* director. As established in its organic act, NSF was empowered to develop and encourage a national policy for the promotion of basic research and science education, to support basic research in the mathematical, physical, medical, biological, engineering, and “other” (e.g., social) sciences. Section 3(b) addressed the geographic distribution issue obliquely, stating that it

> shall be one of the objectives of the Foundation to strengthen basic research and education in the sciences, including independent research by individuals, throughout the United States, including its Territories and possessions, and to avoid undue concentration of such research and education.[65]

As with prior versions of the bill, NSF’s organic act specifically authorized the foundation to provide for scholarships and fellowships, to foster information exchange among scientists in the U.S. and abroad, to establish commissions, to act as a central clearinghouse for information about scientific and technological personnel, and to establish research divisions. With respect to the patent issues, P.L. 81-507 left these questions to the NSF to decide through the contract process.[66] With one notable exception, Congress did not pass another NSF authorization act for the next 15 years. (NSF’s organic act provided the foundation with $500,000 in FY1951 and $15,000,000 annually thereafter. Congress amended the act in 1953 to provide “such sums as may be necessary” (P.L. 83-223).)

The next major reauthorization of the NSF organic act came in 1968.[67] In 1965, the House Committee on Science and Astronautics, Subcommittee on Science, Research and Development (chaired by Representative Emilio Daddario) undertook an extensive, three-year examination of the foundation’s activities and legal authority. Some historians assert that renewed interest in the NSF organic act stemmed from concern about U.S. science policy post-*Sputnik*.[68] The result of the Daddario committee’s work was P.L. 90-407 (An Act to Amend the National Science Foundation Act of 1950). P.L. 90-407 made several critical changes to the NSF organic act that harkened back to the establishment debates of the 1940s. In particular, the act expressly authorized NSF activities in the social sciences and it specifically authorized support for applied research.

P.L. 90-407 also changed NSF’s authorization cycle. The 1968 act repealed the indefinite authorization established by P.L. 83-223 in 1953 and replaced it with an annual authorization. The one-year authorization cycle established by P.L. 90-407 was in place (generally) from FY1969 until FY1989. It was not unchallenged, however. During the late 1970s and early 1980s Congress debated whether to maintain the one-year authorization cycle for NSF. Some Members of Congress preferred tighter oversight and control over the foundation and therefore argued for the one-year authorization.[69]

Other Members asserted that longer authorization cycles would assist in long-range planning, ensure stable funding, and facilitate "sound national science policy and programs."[70] These legislators typically argued for at least two-year authorizations.

Since FY1989 NSF authorization cycles have generally extended beyond a single year. Enacted authorizations for the NSF over the past two decades have typically fluctuated between three and five years. (See **Table A-1**.)

America COMPETES

Since 2007, Congress has included language to reauthorize the NSF in broader bills that, among other things, also authorized scientific research at the Department of Energy's Office of Science and the National Institute of Standards and Technology. Known colloquially as the "COMPETES acts,"[71] these measures authorized FY2008 through FY2013 funding levels for selected federal research accounts, authorized certain federal STEM education programs, and addressed various other policy issues associated with innovation and national competitiveness. NSF provisions in the 2007 and 2010 COMPETES acts included funding authorizations for most major foundation accounts as well as policy provisions authorizing or amending specified policies and programs related to research, STEM education, and broadening participation. Legislators in the 113th Congress have begun considering the reauthorization of major provisions from the COMPETES acts.

Doubling Path

A primary policy question facing the next NSF reauthorization is whether to continue authorizing funding increases for NSF as part of the COMPETES acts "doubling path" policy. Under this policy, Congress and two successive Administrations sought to double—over several years—combined funding for certain federal accounts[72] (including NSF) that fund substantial levels of physical sciences and engineering (PS&E) research.[73] PS&E research is widely believed to contribute to U.S. economic growth and national security by creating the underlying knowledge that supports technological innovation. The COMPETES-authorized PS&E doubling effort followed a successful effort to double funding for medical research at the National Institutes of Health.[74]

As enacted in the 2007 America COMPETS Act, funding authorizations for the targeted accounts set a pace for doubling over approximately seven years (as compared to the FY2006 baseline). However, actual appropriations followed about an 11-year doubling pace during the FY2008 to FY2010 authorization period. The America COMPETES Reauthorization Act of 2010, which authorized funding for the targeted accounts from FY2011 through FY2013, established a doubling pace of about 11 years. (That is, the reauthorization followed a growth rate that was consistent with the growth in actual appropriations during the first authorization period.)

Appropriations to the targeted accounts decreased from the prior year in both FY2011 and FY2013, and as a result, increased the doubling timeframe set by actual appropriations from about 11 years to about 18 years.[75]

The idea of an NSF budget doubling did not originate with the COMPETES acts. President Ronald Reagan proposed a five-year doubling of the NSF budget in January 1987.[76] His FY1988 and FY1989 budget requests sought increases that were consistent with this approach. In October 1988, Congress enacted funding authorizations that sought to double NSF's budget in approximately five years as part of P.L. 100-570 (National Science Foundation Authorization Act of 1988). Actual appropriations to the NSF increased by about 59% during this period.[77] In 2002, Congress passed and President George W. Bush signed, P.L. 107-368 (National Science Foundation Authorization Act of 2002). P.L. 107-368 authorized increases in the NSF budget that were consistent with a five-year doubling. However, the Bush Administration is reported to have objected to the notion of doubling as an arbitrary goal for the NSF, and language referring to doubling was removed from the final bill, though the authorization increases remained.[78] Actual appropriations to the NSF increased by about 22% during the P.L. 107-368 authorization period.[79] President Bush later proposed a doubling similar to that authorized by the COMPETES acts—for example, focused on the targeted accounts, not just NSF—in the 2006 American Competitiveness Initiative.[80]

Viewed in appropriations by decade (e.g., FY1951 to FY1960, FY1960 to FY1970, etc.), the NSF budget doubled (in current dollars) over the course of each of the five decades between the foundation's first budget in FY1951 and FY1990.[81] (See **Table 1**.) Growth slowed from this pace around the turn of the 21st century. Between FY1990 and FY2000, the NSF budget grew by about 88% in current dollars; between FY2000 and FY2010, it grew by about 76% in current dollars.[82]

In inflation-adjusted (constant) dollars, NSF's budget doubled between FY1951 and FY1960, and again between FY1960 and FY1970. The NSF budget has not doubled by decade (in constant dollars) since then. Between FY1970 and FY1980, NSF's budget grew at its lowest constant dollar rate

(16%). Between FY1980 and FY2010, NSF constant dollar funding increased by 38% or more each decade. However, constant dollar funding for NSF was below FY2010 levels in FY2011, FY2012, FY2013, and FY2014, which shows that funding for NSF has not kept pace with inflation this decade.

Table 1. NSF Appropriations by Decade: FY1951 to FY2010 In Millions, Current and Constant (FY2015) Dollars, Rounded

Year	Current ($ millions)	Constant (FY2015 $ millions)
FY1951	0	2
FY1960	153	963
FY1970	440	2,163
FY1980	992	2,500
FY1990	2,082	3,445
FY2000	3,912	5,266
FY2010	6,873	7,471

Source: Excerpted from Table B-1.

Should Congress continue to pursue the COMPETES doubling policy in FY2014 and beyond? Many advocates assert that federal funding for PS&E basic research is inadequate (particularly in light of other countries' investments in R&D) and that more investment is needed to assure U.S. national security and competitiveness.[83] Continuing to provide increased authorizations for the targeted accounts at some (to be determined) doubling rate might signal Congress's continued commitment to these accounts and to the doubling path policy. Further, some analysts have argued that the PS&E doubling policy, although not fully realized, may have protected the targeted accounts from reductions or slower growth during a period of constrained resources.[84] On the other hand, it may be challenging for the scientific community to plan for large or longterm projects without a clearer signal from Congress as to the actual budgetary resources they might receive.[85] Some policy makers who seek general reductions in federal expenditures may object to policies that seek to increase federal spending. However, other observers describe federal funding for basic research (like that funded by the NSF) as the "seed corn" that supports the U.S. economy.[86] These analysts may perceive NSF funding as a vital investment. Another view holds that federal funding for scientific research should be continued, but asserts that dollars should be focused "where links between science and application are well established, to deliver short- to medium-term benefits" rather than on the types of research

NSF typically supports, which may be perceived by some observers as less targeted or less immediately commercially relevant.[87]

STEM Education

Several inventories of the federal STEM education effort have highlighted NSF's important role—both in terms of funding and in the number and breadth of programs—in the federal STEM education portfolio. The NSF is the only federal agency whose primary mission includes supporting education across all fields of science and engineering. As such, funding for STEM education at the NSF impacts not only the agency, but also the entire federal STEM education effort.[88]

The COMPETES acts authorized increased funding for NSF's main education account, Education and Human Resources (EHR), and made various changes to specified NSF STEM education programs. Actual appropriations to EHR have not typically reached COMPETES-authorized levels. Further, Congress reduced enacted funding levels (from the prior year) for EHR in both FY2011 and FY2012. These reductions followed several years of fluctuating funding, as well as changes in the distribution of the foundation budget that reduced funding for EHR as a percentage of the total NSF budget. FY2013 actual funding levels for EHR were close to FY2012 levels.[89]

In addition to funding authorizations, the COMPETES acts authorized and amended some NSF STEM education programs.[90] Among the amended programs were the Graduate Research Fellowship (GRF) program and the Integrative Graduate Research and Education Traineeship (IGERT). First, Section 510 of the America COMPETES Reauthorization Act of 2010 sought to require NSF to treat the GRF and IGERT equally by increasing or decreasing funding for these programs at the same rates. This does not appear to have been implemented. Funding for the IGERT was reduced from the prior year each fiscal year between FY2011 and FY2013 while funding for the GRF increased each year. Second, Section 510 also directed NSF to draw from the RRA account for at least half of the funding it provides to the GRF and IGERT programs. RRA funding for the GRF and IGERT programs was close to 50% in FY2012 and FY2013.

The GRF was established in 1951 and is one of the oldest and most prestigious federal graduate research fellowships. GRF fellows receive a three-year, portable stipend of $32,000 annually and a $12,000 cost-of-education allowance for tuition and fees (paid to their institutions).[91] The IGERT, which began in 1997, is NSF's flagship interdisciplinary training program.[92] IGERT funding is awarded to institutions of higher education, which may utilize

IGERT funding for student support or education research. In FY2013, NSF provided funding for 5,758 GRF fellows[93] and 1,572 IGERT trainees.

After the expiration of major provisions in the 2010 COMPETES reauthorization, President Obama proposed a major restructuring of federal STEM education programs as part of his FY2014 budget request. Under the proposed reorganization, NSF would have become the lead federal agency for undergraduate education and federal fellowships. For a range of reasons, congressional appropriators largely rejected the plan.[94] The Obama Administration released a "fresh" reorganization plan as part of the FY2015 budget request.[95] Congressional debate about the reorganization of the federal STEM education effort is ongoing.

For many policy makers, the prospect of a reorganized federal STEM education effort raises the question, "to what end?" The America COMPETES Reauthorization Act of 2010 directed the National Science and Technology Council (NSTC) to develop a strategy for federal STEM education programs, including those at NSF. That strategy was published in May 2013, after release of the Administration's FY2014 proposed reorganization of federal STEM education programs.[96] Some policy makers perceived the strategy as insufficiently independent from the FY2014 proposed reorganization;[97] while others perceived it as a starting place for a new conversation about the federal STEM education portfolio in lieu of the proposed reorganization.[98] Policy makers' interest in the form and function of the federal STEM education portfolio, as well as the character of NSF's role within that portfolio, continues.

Reauthorization Activity in the 113th Congress

Major provisions of the COMPETES acts, including provisions authorizing funding for the National Science Foundation, expired in FY2013. The House Committee on Science, Space, and Technology has begun the process of reauthorizing certain COMPETES provisions. On March 13, 2014, the Subcommittee on Research and Technology marked up and approved by voice vote the Frontiers in Innovation, Research, Science, and Technology (FIRST) Act (H.R. 4186).[99] The full committee marked up the FIRST Act on May 21, 2014, and voted on amendments on May 28, 2014. The ranking Member of the House Committee on Science, Space, and Technology has introduced an alternative bill, H.R. 4159 (America COMPETES Reauthorization Act of 2014). H.R. 4159 had not been marked up as of the date of this report. H.R. 4186 and H.R. 4159 differ from each other, and from the COMPETES acts, in significant ways. These differences, however, are beyond the scope of this report.

BUDGET AND APPROPRIATIONS

Like other federal agencies, NSF's annual budget requests to Congress provide insight into foundation activities and priorities.[100] A summary of NSF's three most recent budget requests (FY2013, FY2014, and FY2015), and associated appropriations activity, follows. **Table B-1** provides NSF authorizations, budget requests, and actual appropriations in current and constant (inflation-adjusted) dollars from FY1951 to FY2015. **Table B-2** provides FY2003 to FY2013 NSF obligations by major account.

NSF adopted its current appropriations account structure in FY2003. In general, NSF's major accounts have been comparable since then.[101] NSF has six major appropriations accounts: Research and Related Activities (RRA), Education and Human Resources (EHR), Major Research Equipment and Facilities Construction (MREFC), Agency Operations and Award Management (AOAM), National Science Board (NSB), and the Office of the Inspector General (OIG). The majority of NSF's primary mission activities are funded through RRA and EHR.

FY2015

For FY2015, the Obama Administration seeks $7.255 billion in funding for the NSF. This amount is $83 million (1.2%) over the FY2014 enacted level of $7.172 billion. The request holds funding levels for RRA and MREFC essentially constant while seeking a 5.1% increase ($43 million) for EHR as well as a 13.5% increase ($40 million) for AOAM. Most of the new AOAM funding would apply toward a new NSF headquarters. NSF's FY2015 budget request to Congress highlights five initiatives that were also foundation priorities in FY2014 and FY2013: Cognitive Science and Neuroscience ($29 million); Cyber-enabled Materials, Manufacturing, and Smart Systems ($213 million); Cyberinfrastructure Framework for 21st Century Science, Engineering, and Education ($125 million); Science, Engineering, and Education for Sustainability ($139 million); and Secure and Trustworthy Cyberspace ($100 million).[102] The FY2015 NSF budget request incorporates

STEM education programs changes in accordance with the Administration's revised FY2015 government-wide reorganization of federal STEM education programs.[103] In addition to NSF's regular budget request, the Administration seeks $552 million in funding for NSF through the proposed Opportunity, Growth, and Security Initiative.[104]

Congress had not enacted legislation authorizing FY2015 funding levels for NSF when the President released the FY2015 budget request.

FY2014

FY2014 enacted funding for NSF is $7.172 billion. This amount is $270 million (3.9%) more than NSF's FY2013 actual funding level of $6.902 billion. Most of the $270 million increase ($250 million) went to RRA. FY2014 enacted funding for NSF's six major accounts is $5.809 billion for RRA (including $158 million for EPSCoR), $847 million for EHR, $200 million for MREFC, $298 million for AOAM, $4 million for NSB, and $14 million for OIG.[105] Congress had not enacted legislation authorizing funding for the NSF in FY2014 when FY2014 appropriations were enacted.

The Obama Administration initially sought $7.626 billion in funding for the NSF in FY2014. NSF's FY2014 budget request to Congress noted that its overarching priorities for FY2014 would include six programs: Cyber-enabled Materials, Manufacturing, and Smart Systems; Cyberinfrastructure Framework for 21st Century Science, Engineering, and Education; NSF Innovation Corps; Integrated NSF Support Promoting Interdisciplinary Research and Education; Science, Engineering, and Education for Sustainability; and Secure and Trustworthy Cyberspace.[106] The FY2014 NSF budget request also incorporated several changes to the foundation's STEM education programs in accordance with the Administration's proposed FY2014 government-wide reorganization of federal STEM education programs.[107]

The House and Senate Committees on Appropriations recommended $6.995 billion and $7.426 billion, respectively, for NSF in FY2014. Both committees initially rejected the Administration's proposed changes to the federal STEM education effort, including changes to NSF programs. The final FY2014 appropriations agreement reiterated this objection. The appropriations committees initially disagreed on funding for the Large Synoptic Survey Telescope (LSST) in the MREFC account—the Senate Committee on Appropriations sought to fund the new project, the House Committee on Appropriations would not. The final agreement provided some of the requested funding for the LSST and encouraged the foundation to seek permission to transfer funds from other accounts if the amount appropriated was insufficient.[108]

FY2013

FY2013 actual funding for NSF was $6.902 billion. This amount was $199.4 million (3.5%) less than NSF's FY2012 actual funding level of $7.105 billion.[109] FY2013 actual funding levels for NSF's six major accounts were $5.559 billion (RRA), $835 million (EHR), $197 million (MREFC), $294 million (AOAM), $4 million (NSB), and $14 million (OIG). The America COMPETES Reauthorization Act of 2010 (P.L. 111-358) authorized $8.300 billion in funding at the NSF in FY2013.

The Obama Administration sought $7.373 billion for the NSF in FY2013. NSF's FY2013 budget request sought to continue increasing NSF's budget in accordance with the doubling path policy and emphasized "OneNSF Framework" priorities. NSF's FY2013 budget documents asserted that the six OneNSF Framework priority programs would "create new knowledge, stimulate discovery, address complex societal problems, and promote national prosperity."[110]

The full House and Senate Committee on Appropriations agreed on identical funding levels for five of six major NSF accounts in FY2013. The primary difference between the two proposals was in the main research account (RRA). The House proposed $5.943 billion for RRA in FY2013 while the Senate Committee on Appropriations recommended $5.883 billion. This difference in funding for RRA lead to an equivalent difference in House and Senate topline recommendations for NSF, which were $7.333 billion and $7.273 billion, respectively. FY2013 House and Senate recommendations for NSF's other five major accounts were $875.6 million (EHR), $196.2 million (MREFC), $299.4 million (AOAM), $4.4 million (NSB), and $14.2 million (OIG). These amounts equaled the Administration's requested funding levels.

CONCLUDING OBSERVATIONS

The National Science Foundation plays a key role in the federal research and development, as well as STEM education, portfolios. It is a primary source of support for basic research in fields that many analysts cite as key to future competitiveness, such as the physical sciences, mathematics, and computer science. It is also a primary source for federal STEM education funding. Yet, the foundation differs from many other federal agencies in a number of key ways. It was established outside of the central core of executive agencies rather than under the direct control of the President. It focuses on fundamental research across a variety of scientific and technological fields rather than on the specific mission needs of the federal government; and it is the only federal agency whose primary mission includes STEM education. These differences underpin much of the policy conversation about the NSF; and, as a practical matter, offer both benefits and barriers to legislators seeking to apply NSF's various assets toward specific national goals.

APPENDIX A. NSF AUTHORIZATIONS ACTS

Table A-1. Selected NSF Authorizations Acts FY1951-FY2013

Public Law	Bill Number	From	To
P.L. 81-507	S. 247	FY1951	FY1952
P.L. 83-223	S. 977	FY1953	not defined
authorizations not defined		FY1954	FY1968
P.L. 90-407	H.R. 5404	FY1969	FY1969
P.L. 91-120	S. 1857	FY1970	FY1970
P.L. 91-356	H.R. 16595	FY1971	FY1971
P.L. 92-86	H.R. 7960	FY1972	FY1972
P.L. 92-372	H.R. 14108	FY1973	FY1973
P.L. 93-96	H.R. 8510	FY1974	FY1974
P.L. 93-413	H.R. 13999	FY1975	FY1975
P.L. 94-86	H.R. 4723	FY1976	FY1976
P.L. 94-471	H.R. 12566	FY1977	FY1977
P.L. 95-99	H.R. 4991	FY1978	FY1978
P.L. 95-434	H.R. 11400	FY1979	FY1979
P.L. 96-44	H.R. 2729	FY1980	FY1980
P.L. 96-516	S. 568	FY1981	FY1981
some authorizations are introduced, none become law		FY1982	FY1985
P.L. 99-159	H.R. 1210	FY1986	FY1986
P.L. 99-383	H.R. 4184	FY1987	FY1987
some authorizations are introduced, none become law		FY1988	FY1988
P.L. 100-570	H.R. 4418	FY1989	FY1993

Table A-1. (Continued)

Public Law	Bill Number	From	To
some authorizations are introduced, none become law		FY1994	FY1997
P.L. 105-207	H.R. 1273	FY1998	FY2000
some authorizations are introduced, none become law		FY2001	FY2001
P.L. 107-368	H.R. 4664	FY2003	FY2007
P.L. 110-69	H.R. 2272	FY2008	FY2010
P.L. 111-358	H.R. 5116	FY2011	FY2013

Source: Congressional Research Service, based on information from the Legislative Information System and Proquest Congressional.

Notes: This table includes a list of major NSF authorization acts as per a CRS search of historical legislative databases. The list of authorization acts has been reviewed by National Science Board legal counsel, who affirmed its apparent completeness. In addition to the above-listed authorization acts, other laws have also amended various parts of the NSF code.

APPENDIX B. NSF FUNDING HISTORY

Table B-1. NSF Authorizations, Budget Requests, and Appropriations FY1951 to FY2015 In Millions, Current and Constant (FY2015) Dollars, Rounded

Fiscal Year	Current ($ millions)			Constant (FY2015 $ millions)		
	Authorization	Request	Appropriation	Authorization	Request	Appropriations
1951	such sums	-	0	n/a	-	2
1952	such sums	14	4	n/a	103	26
1953	such sums	15	5	n/a	109	34
1954	such sums	15	8	n/a	108	57
1955	such sums	14	14	n/a	100	101
1956	such sums	31	53	n/a	215	367
1957	such sums	41	40	n/a	276	267
1958	such sums	65	52	n/a	422	336
1959	such sums	140	138	n/a	894	878
1960	such sums	160	153	n/a	1,010	963
1961	such sums	190	176	n/a	1,180	1,092
1962	such sums	210	263	n/a	1,292	1,619
1963	such sums	358	323	n/a	2,175	1,959
1964	such sums	589	353	n/a	3,535	2,118
1965	such sums	488	420	n/a	2,877	2,479

Fiscal Year	Current ($ millions)			Constant (FY2015 $ millions)		
	Authorization	Request	Appropriation	Authorization	Request	Appropriations
1966	such sums	530	480	n/a	3,061	2,771
1967	such sums	525	481	n/a	2,942	2,695
1968	such sums	526	495	n/a	2,850	2,682
1969	525	500	400	2,719	2,590	2,072
1970	478	500	440	2,348	2,458	2,163
1971	538	513	513	2,516	2,400	2,400
1972	653	622	622	2,913	2,777	2,777
1973	697	653	649	2,982	2,795	2,778
1974	633	583	579	2,528	2,328	2,315
1975	808	672	764	2,925	2,435	2,768
1976	787	755	715	2,666	2,559	2,423
1977	811	802	776	2,562	2,535	2,452
1978	879	944	863	2,603	2,795	2,555
1979	930	934	911	2,549	2,560	2,497
1980	1,002	1,006	992	2,525	2,536	2,500
1981	1,115	1,148	1,025	2,559	2,636	2,354
1982	n/a	1,354	1,039	n/a	2,909	2,232
1983	n/a	1,073	1,094	n/a	2,208	2,251
1984	n/a	1,292	1,341	n/a	2,569	2,665
1985	n/a	1,502	1,502	n/a	2,890	2,889
1986	1,517	1,569	1,524	2,854	2,952	2,867
1987	1,685	1,686	1,623	3,101	3,102	2,986
1988	n/a	1,893	1,717	n/a	3,375	3,061
1989	2,050	2,050	1,923	3,515	3,515	3,296
1990	2,388	2,149	2,082	3,951	3,556	3,445
1991	2,782	2,485	2,316	4,445	3,971	3,701
1992	3,245	2,742	2,571	5,061	4,277	4,009
1993	3,505	3,037	2,734	5,340	4,627	4,165
1994	n/a	2,753	2,983	n/a	4,105	4,447
1995	n/a	3,200	3,264	n/a	4,672	4,765
1996	n/a	3,360	3,220	n/a	4,816	4,615
1997	n/a	3,325	3,270	n/a	4,683	4,606
1998	3,506	3,367	3,431	4,877	4,684	4,773
1999	3,773	3,773	3,676	5,184	5,184	5,050
2000	3,886	3,921	3,912	5,231	5,278	5,266
2001	n/a	4,572	4,431	n/a	6,009	5,823
2002	n/a	4,473	4,823	n/a	5,785	6,239
2003	5,536	5,036	5,323	7,027	6,391	6,756
2004	6,391	5,481	5,589	7,915	6,789	6,922
2005	7,378	5,745	5,482	8,860	6,899	6,584

Table B-1. (Continued)

Fiscal Year	Current ($ millions)			Constant (FY2015 $ millions)		
	Authorization	Request	Appropriation	Authorization	Request	Appropriations
2006	8,520	5,605	5,589	9,909	6,519	6,500
2007	9,839	6,020	5,890	11,142	6,817	6,670
2008	6,600	6,429	6,125	7,323	7,133	6,796
2009	7,326	6,854	6,494	8,034	7,516	7,121
2010	8,132	7,045	6,873	8,841	7,659	7,471
2011	7,424	7,424	6,806	7,917	7,917	7,257
2012	7,800	7,767	7,033	8,174	8,140	7,371
2013	8,300	7,373	6,884	8,569	7,612	7,107
2014	n/a	7,626	7,172	n/a	7,757	7,295
2015	n/a	7,255	-	n/a	7,255	-

Source: Funding data in the "Authorization" columns are from selected FY1951 to FY2013 NSF authorization acts, as provided in Table A-1. Funding data in the "Request" and "Appropriations" columns are from National Science Foundation, Budget Internet Information System, "NSF Requests and Appropriations History," NSF.gov, accessed March 19, 2014, http://dellweb.bfa.nsf.gov/NSFRqst AppropHist/ NSFRequestsandAppropriationsHistory.pdf. To calculate constant dollars, CRS applied the Gross Domestic Product, Chained Price Index (adjusted to reflect FY2015 dollars) found in Office of Management and Budget, "Table 10.1," *Historical Tables*, accessed on May 6, 2014, available at http://www.whitehouse.gov/sites/default/files/ omb/budget/fy2015/assets/hist10z1.xls.

Notes: Totals may not add due to rounding. As per communication between CRS and NSF dated March 20, 2014, the "appropriation" column shows funding provided in annual appropriations acts plus adjustments required in those acts, other laws, and committee reports, etc. Adjustments include rescissions, sequestration, funding transfers across NSF accounts, supplemental appropriations (not including American Recovery and Reinvestment Act, P.L. 111-5, funding in FY2009), and other changes. The resulting amounts most closely align with NSF's approved Current Plans.

Table B-2. NSF Obligations by Major Account: FY2003-FY2013
In Millions, Current Dollars, Rounded

Fiscal Year	RRA	EHR	MREFC	AOAM	NSB	OIG	NSF Total
2003	$4,144	$846	$179	$189	$3	$9	$5,369
2004	$4,388	$850	$184	$219	$2	$9	$5,652
2005	$4,328	$750	$165	$223	$4	$10	$5,481
2006	$4,449	$700	$234	$247	$4	$11	$5,646
2007	$4,758	$696	$166	$248	$4	$12	$5,884
2008	$4,853	$766	$167	$282	$4	$12	$6,084
2009	$5,152	$846	$161	$294	$4	$12	$6,469
2010	$5,615	$873	$166	$300	$4	$14	$6,972
2011	$5,608	$861	$125	$299	$4	$14	$6,913
2012	$5,758	$831	$198	$299	$4	$14	$7,105
2013	$5,559	$835	$196	$294	$4	$14	$6,902

Source: National Science Foundation annual budget requests to Congress from FY2005 to FY2015.

Notes: NSF adopted its current appropriations account structure in 2003. CRS adjusted FY2003 to FY2007 RRA and EHR obligations data to reflect the transfer of the EPSCoR program between these accounts in FY2008. This table treats EPSCoR as a research account program for all years in the data set. Does not include American Recovery and Reinvestment Act (ARRA, P.L. 111-5) funding.

End Notes

[1] P.L. 81-507.

[2] See David E. Lewis and Jennifer L. Selin, *Sourcebook of United States Executive Agencies*, 1st ed. (Washington, DC: Administrative Conference of the United States, March 2013), p. 54; J. Merton England, "National Science Foundation," in *Government Agencies*, ed. Donald R. Whitnah (Westport, CT: Greenwood Press, 1983), pp. 367-372; and Harold Seidman, "A Typology of Government," in *Federal Reorganization: What Have We Learned*, ed. Peter Szanton (Chatham, NJ: Chatham House Publishers, Inc. 1981), pp. 43-44.

[3] The NSF director must also be confirmed by the Senate.

[4] 42 U.S.C. 1861, §2.

[5] For more information about independent agencies—including rationales for, historical origin of, and accountability in—see the section titled, "Background and Context" in CRS Report R43391, *Independence of Federal Financial Regulators*, by Henry B. Hogue, Marc Labonte, and Baird Webel.

[6] England, p. 367.

[7] Lewis and Selin, p. 59.

[8] More information about NSF leadership and staff may be found in Stephen Horn (panel chair), et al., *National Science Foundation: Governance and Management for the Future*, National

Academy of Public Administration, April 2004, p. xv, at http://www.napawash.org/2004 /1539-national-science-foundation-governance-and-management-for-thefuture.html.

[9] National Science Foundation, "About the National Science Foundation: Who We Are," *National Science Foundation Website*, accessed February 7, 2014, at http://www.nsf.gov/about/who.jsp.

[10] National Science Board, "About the NSB," *National Science Board Website*, accessed February 7, 2014, http://www.nsf.gov/nsb/about/; and National Science Foundation, "Introduction," *Proposal and Award Policies and Procedures Guide* (NSF 13-1), January 13, 2013, at http://www.nsf.gov/pubs/policydocs/pappguide/nsf13001/ index.jsp. See also, the section titled "Grant-Making" in this report.

[11] Board members must be "eminent in the fields of the basic, medical, or social sciences, engineering, agriculture, education, research management or public affairs...." (42 U.S.C. 1863(c)(1)).

[12] National Science Foundation, "List of NSF Directors, 1950-Present," *National Science Foundation Website*, accessed February 7, 2013, at http://www.nsf.gov/od/nsf-director-list/nsf-directors.jsp.

[13] Presidential Appointment Efficiency and Streamlining Act of 2011 (P.L. 112-166, §2 (s)).

[14] A list of NSF deputy directors is available at http://www.nsf.gov/od/nsf-director-list/nsf-deputys.jsp.

[15] For example, see Jeffrey Mervis, "Special Report: Can NSF Put the Right Spin on Rotators? Part 1," *Science Insider*, October 10, 2013, http://news.sciencemag.org/policy/2013/10 /special-report-can-nsf-put-right-spin-rotators-part-1; and Jeffrey Mervis, "Special Report: Can NSF Put the Right Spin on Rotators? Part 2," *Science Insider*, October 24, 2013, at http://news.sciencemag.org/people-events/2013/10/special-report-can-nsf-put-right-spin-rotators-part-2.

[16] Stephen Horn (panel chair), et al., *National Science Foundation: Governance and Management for the Future*, National Academy of Public Administration, April 2004, p. xv, at http://www.napawash.org/2004/1539-nationalscience-foundation-governance-and-management-for-the-future.html.

[17] Memorandum from Assistant Inspector General for Audit Dr. Brett M. Baker, National Science Foundation, Office of Inspector General to Deputy Director Dr. Cora B. Marrett, National Science Foundation, dated March 20, 2013, at http://www.nsf.gov/oig/13-2-006.pdf.

[18] OMB Circular A-11, Schedule C, defines basic research as "systematic study directed toward fuller knowledge or understanding of the fundamental aspects of phenomena and of observable facts without specific applications towards processes or products in mind. Basic research, however, may include activities with broad applications in mind." Basic research differs from applied research, which is "systematic study to gain knowledge or understanding necessary to determine the means by which a recognized and specific need may be met;" and from development, which is the "systematic application of knowledge or understanding, directed toward the production of useful materials, devices, and systems or methods, including design, development, and improvement of prototypes and new processes to meet specific requirements." See, Office of Management and Budget, "Character Classification (Schedule C)," *OMB Circular A-11* (2013), at http://www.whitehouse.gov /sites/default/files/omb/assets/a11_current_year/s84.pdf.

[19] Based on preliminary FY2012 data from Tables 29 and 22 of the National Science Foundation, National Center for Science and Engineering Statistics, Federal Funds for Research and Development: Fiscal Years 2010-12, NSF 13-326 (July 2013).

[20] NSF's organizational chart is available at http://www.nsf.gov/staff/orglist.jsp. In addition to the research and education directorates, NSF also has two administrative offices: the Office of Budget, Finance, & Award Management and the Office of Information & Resource Management.

[21] National Science Foundation, Directorate for Biological Sciences, "About Biological Sciences," *National Science Foundation Website*, accessed February 11, 2014, at http://www.nsf.gov/bio/about.jsp. See also, National Science Foundation, *Understanding Life: Biological Sciences*, NSF 14-804 (March 2014), at https://www.nsf.gov/about/congress/reports/bio_research.pdf.

[22] National Science Foundation, Directorate for Computer and Information Science and Engineering, "CISE—About," *National Science Foundation Website*, accessed February 11, 2014, at http://www.nsf.gov/cise/about.jsp. See also, National Science Foundation, *Enhancing Our Lives through Computing: Computer & Information Science & Engineering*, NSF 14-807 (February 2014), at https://www.nsf.gov/about /congress /reports/cise_research.pdf.

[23] National Science Foundation, Directorate for Education and Human Resources, "About Education and Human Resources," *National Science Foundation Website*, accessed February 11, 2014, at http://www.nsf.gov/ehr/about.jsp. See also, National Science Foundation, *Inspiring STEM Learning: Education & Human Resources*, NSF 12-800 (September 2013), at https://www.nsf.gov/about/congress/reports/ehr_research.pdf.

[24] National Science Foundation, Directorate for Engineering, "General Information about ENG," *National Science Foundation Website*, accessed March 25, 2014, at http://www.nsf.gov/eng/about.jsp. See also, National Science Foundation, *Making Future Technologies Possible: Engineering*, NSF14-808 (April 2014), at https://www.nsf.gov/about/congress/reports/eng_research.pdf.

[25] National Science Foundation, Directorate for Geosciences, "About GEO," *National Science Foundation Website*, accessed February 11, 2014, at http://www.nsf.gov/geo/about.jsp. See also, National Science Foundation, *Unraveling Earth's Complexity: Geosciences*, NSF 13-801 (September 2013), at https://www.nsf.gov/about/congress/reports/ geo_research.pdf.

[26] National Science Foundation, Directorate for Mathematical and Physical Sciences, "About Directorate for Mathematical and Physical Sciences," *National Science Foundation Website*, accessed February 11, 2014, at http://www.nsf.gov/mps/about.jsp. See also, National Science Foundation, *Enriching the Language of Discovery: Mathematical & Physical Sciences*, NSF 14-805 (March 2014), https://www.nsf.gov/about/congress/reports/mps_research.pdf.

[27] National Science Foundation, *FY2015 Budget Request to Congress*, March 10, 2014, p. SBE-1. See also, National Science Foundation, Exploring What Makes Us Human: Social, Behavioral & Economic Sciences, NSF 14-803 (March 2014), at https://www.nsf.gov/about /congress/reports/sbe_research_2.pdf.

[28] Rep. Lamar Smith, "Commerce, Justice, Science, and Related Agencies Appropriations Act, 2015," remarks in the House, *Congressional Record*, daily edition, vol. 160 (May 29, 2014), pp. H4958.

[29] National Science Foundation, *Bringing People into Focus: How Social, Behavioral and Economic Research Addresses National Challenges*, at https://www.nsf.gov/about /congress/reports/sbe_research.pdf. See also, Jameson M. Wetmore, "The Value of the Social Sciences for Maximizing Public Benefits of Engineering," *The Bridge: Linking Engineering and Society*, vol. 42, no. 3 (Fall 2012), p. 41.

[30] Smith, ibid., pp. H4957-H4960.

[31] For example, see Sen. Tom Coburn, M.D., "Dr. Coburn Offers Amendments to 587-Page, $1 Trillion Continuing Resolution to Fund the Government," press release, March 13, 2103, at http://www.coburn.senate.gov/public/ index.cfm/rightnow?ContentRecord_id=1167a7ba-31a0-4873-b5c9-a2acd77eb72b.

[32] As adopted, Section 543 prohibited NSF from using FY2013 appropriations to carry out the functions of the Political Science Program, except for projects that promote U.S. national security and economic interests. Section 543 further required the NSF to certify that political science projects met this requirement, to publish the reason for each certification on its website, and allowed NSF to use any unobligated balances from the Political Science Program for other research.

[33] For example, see American Political Science Association, "Senate Delivers a Devastating Blow to the Integrity of the Scientific Process at the National Science Foundation," press release, March 20, 2013, at http://www.prnewswire.com/ news-releases/senate-delivers-a-devastating-blow-to-the-integrity-of-the-scientific-process-at-the-national-science-foundation-199221111.html.

[34] NSF's *Proposal and Award Policies and Procedures Guide* (PAPPG) describes the foundation's grant-making process and provides guidance to potential applicants.

[35] PAPPG, "Introduction."

[36] One exception to this rule is the RAPID (Grants for Rapid Response Research) funding mechanism. RAPID grants may be used for "proposals having a severe urgency with regard to availability of, or access to data, facilities or specialized equipment, including quick-response research on natural or anthropogenic disasters and similar unanticipated events." Only internal merit review is required for RAPID grants. PAPPG, p. II-22.

[37] More information on the proposal preparation and submission phase is available at http://www.nsf.gov/bfa/dias/ policy/merit_review/phase1.jsp.

[38] More information on the proposal review and processing phase is available at http://www.nsf.gov/bfa/dias/policy/ merit_review/phase2.jsp.

[39] More information on the award processing phase is available at http://www.nsf.gov/bfa /dias/policy/merit_review/ phase3.jsp.

[40] Peer review can happen in a number of ways. Reviewers may be sought out on an individual basis (also known as ad hoc review) or may participate in in-person or virtual panels. While a minimum of three reviewers is required, more may participate.

[41] In addition to these criteria, NSF solicitations may include additional criteria that meet the specific objectives of programs or activities.

[42] More information about NSF's merit review process and criteria is available at the NSF Merit Review FAQ webpage, available at http://www.nsf.gov/bfa/dias/policy/merit_review /facts.jsp; and in "Chapter III" of the PAPPG, available at http://www.nsf.gov/pubs /policydocs/pappguide/nsf13001/gpg_3.jsp#IIIA2a.

[43] National Science Foundation, "Phase II: Proposal Review and Processing," *National Science Foundation Website*, accessed February 11, 2014, http://www.nsf.gov/bfa/dias/policy /merit_review/phase2.jsp.

[44] The National Science Board resolution, "Delegation of Award-Approval Authority to the Director," (NSB-11-2) establishes the conditions under which the NSB delegates its authority to approve NSF awards to the NSF director. Section (2)(B) states that the director may not make an award that "involves sensitive political or policy issues" without Board approval.

[45] The NSB Delegation of Award-Approval Authority to the Director" (Ibid. Section (1)) states that the NSF director "may make no award involving an anticipated average annual amount

of the greater of either 1 percent of more of the awarding Directorate's or Office's prior year current plan *or* [emphasis original] 0.1 percent or more of the prior year total NSF budget without the prior approval of the National Science Board."

[46] Contracts are used to acquire products, services, and studies (e.g., program evaluations).

[47] "Total academic R&D" includes funding from all sources. However, "the federal government provided the majority of the S&E R&D funds that public and private institutions spent on R&D in FY 2012 (just under 60% and just over 70%, respectively)." National Science Board, *Science and Engineering Indicators 2014*, NSB-14-01 (Arlington, VA: National Science Foundation, 2014), p. 5-16 – 5-17.

[48] Initial funding for EPSCoR was provided in P.L. 95-392 (Department of Housing and Urban Development-Independent Agencies Appropriation Act, 1979). More information about EPSCoR is available at http://www.nsf.gov/ div/index.jsp?org=EPSC.

[49] National Science Foundation, EPSCoR Interagency Coordinating Committee, "Eligible EPSCoR Jurisdictions by Agency," fact sheet, last updated May 17, 2012, at http://www.nsf.gov/od/iia/programs/epscor/EICC.pdf.

[50] This paragraph adapted from CRS Report RL30930, *U.S. National Science Foundation: Experimental Program to Stimulate Competitive Research (EPSCoR)*, by Christine M. Matthews (retired).

[51] Lambright, W. Henry, Syracuse University, Paper prepared for the American Association for the Advancement of Science, Workshop on Academic Research Competitiveness, Coeur d'Alene, Idaho, *Building State Science: The EPSCoR Experience*, October 1-3, 1999, at http://www.aaas.org/spp/rcp/epscor/lambright.htm, p.4.

[52] National Academy of Sciences, National Academy of Engineering, and Institute of Medicine, Policy and Global Affairs, Committee on Science, Engineering, and Public Policy, Committee to Evaluate the Experimental Program to Stimulate Competitive Research (EPSCoR) and Similar Federal Agency Programs, *The Experimental Program to Stimulate Competitive Research* (Washington, DC: The National Academies Press, 2013).

[53] Readers who have seen the 1997 film *Contact* may also be familiar with the NSF-funded Very Large Array in Soccorro, New Mexico. National Science Foundation, "Radio Telescopes in the New Movie 'Contact' Dish Up Real Science," press release (97-049), July 10, 1997, available at http://www.nsf.gov/news/news_summ.jsp?cntn_id= 102822.

[54] National Academy of Sciences, National Academy of Engineering, and Institute of Medicine, Committee on Underrepresented Groups and the Expansion of the Science and Engineering Workforce Pipeline, Committee on Science, Engineering, and Public Policy, Policy and Global Affairs, *Expanding Underrepresented Minority Participation: America's Science and Technology Talent at the Crossroads*, National Academies Press, 2011, at http://www.nap.edu/openbook.php?record_id=12984.

[55] Historical narratives about the founding of the NSF typically focus on Sen. Kilgore and Bush but the 79th Congress considered several bills focused on the question of post-war scientific research. See U.S. Senate, Committee on Military Affairs, Subcommittee on War Mobilization, *Legislative Proposals for the Promotion of Science: The Texts of Five Bills and Excerpts from Reports*, subcommittee print, 79th Cong., 1st sess., August 1945.

[56] President Franklin Delano Roosevelt established the OSRD as an independent agency within the Office of Emergency Management (Executive Order 8807). More information about OSRD, as well as holdings, is available on the Library of Congress website at http://www.loc.gov/rr/scitech/trs/trsosrd.html.

[57] U.S. Congress, House Committee on Science and Technology, Task Force on Science Policy, "A History of Science Policy in the United States, 1940-1985," *Science Policy Study*

Background Report No. 1, 99th Cong., 2nd sess., September 1986 (Washington, DC: GPO, 1986), p. 21-27; and, George T. Mazuzan, *National Science Foundation: A Brief History*, NSF 88-16 (Washington, DC: National Science Foundation, 1988).

[58] At the time, most stakeholders agreed with the general concept of a publically funded scientific research foundation. One exception was Frank B. Jewett, then president of the National Academy of Sciences. Jewett expressed concern about unwanted government control and interference in science and preferred private sources of funding. See J. Merton England, *A Patron for Pure Science: The National Science Foundation's Formative Years, 1945-57* (Washington, DC: NSF, 1982), pp. 35-36.

[59] S. 1297 and S. 1285 differed with respect to the roles and authorities assigned to the director and board. S. 1297 gave most of the power to the director (with the board in an advisory capacity); while S. 1285 put most of the authority in the hands of the board, who appointed the director.

[60] U.S. Congress, House Committee on Science and Technology, Task Force on Science Policy, "A History of Science Policy in the United States, 1940-1985," *Science Policy Study Background Report No. 1*, 99th Cong., 2nd sess., September 1986 (Washington, DC: GPO, 1986), p. 21-27; and, J. Merton England, *A Patron for Pure Science: The National Science Foundation's Formative Years, 1945-57* (Washington, DC: NSF, 1982).

[61] S. 1297, Title IV, Section 402 (a) as published in U.S. Senate, Committee on Military Affairs, Subcommittee on War Mobilization, *Legislative Proposals for the Promotion of Science: The Texts of Five Bills and Excerpts from Reports*, subcommittee print, 79th Cong., 1st sess., August 1945.

[62] Harry S. Truman Library & Museum, Public Papers of the Presidents, Harry S. Truman 1945-1953, "169. Memorandum of Disapproval of the National Science Foundation Bill," August 6, 1947, at http://www.trumanlibrary.org/publicpapers/index.php?pid=1918.

[63] S. 247 (National Science Foundation Act of 1950).

[64] Harry S. Truman Library & Museum, Public Papers of the Presidents, Harry S. Truman 1945-1953, "120. Statement by the President Upon Signing Bill Creating the National Science Foundation," May 10, 1950, at http://trumanlibrary.org/publicpapers/index.php?pid=743..

[65] P.L. 81-507, Section 3(b).

[66] For a broader treatment of federal patent issues, please see CRS Report R42014, *The Leahy-Smith America Invents Act: Innovation Issues*, by John R. Thomas.

[67] Although not a reauthorization act per se, in 1962 President John F. Kennedy signed "Reorganization Plan No. 2 of 1962," which established the Office of Science and Technology (OST) within the Executive Office of the President. The plan transferred authority for national science policy making from NSF to OST and made other changes within NSF. Congress had the power to disapprove of this plan, but did not do so, and thereby facilitated implementation. (For more information about the reorganization process, see CRS Report R42852, *Presidential Reorganization Authority: History, Recent Initiatives, and Options for Congress*, by Henry B. Hogue.)

[68] Many analysts and historians consider the Soviet Union's launch of *Sputnik*, the world's first artificial satellite, a watershed moment in U.S. science (and science education) policy history. See CRS Report RL34263, *U.S. Civilian Space Policy Priorities: Reflections 50 Years After Sputnik*, by Deborah D. Stine.

[69] Ken Hechler, *Toward the Endless Frontier: History of the Committee on Science and Technology, 1959-79* (Washington, D.C.: U.S. House of Representatives/GPO, 1980), pp. 537-538.

[70] S.Rept. 95-851, pp. 22-23.

[71] America COMPETES Act (P.L. 110-69) and America COMPETES Reauthorization Act of 2010 (P.L. 111-358).

[72] The targeted accounts included the NSF, the Department of Energy's Office of Science, and the Scientific and Technical Research and Services (STRS) and Construction of Research Facilities (CRF) accounts at the National Institute of Standards and Technology (NIST).

[73] The largest funder of research in engineering is the Department of Defense. The National Aeronautics and Space Administration (NASA) also emphasizes engineering and the physical sciences research. See National Science Board, *Science and Engineering Indicators 2014*, NSB-14-01 (Arlington, VA: National Science Foundation, 2014), p. 4-38.

[74] For more information about the NIH doubling, see CRS Report R43341, *A History of NIH Funding*, by Judith A. Johnson.

[75] For more information about the PS&E doubling effort, see CRS Report R41951, *An Analysis of Efforts to Double Federal Funding for Physical Sciences and Engineering Research*, by John F. Sargent Jr.

[76] President Ronald Reagan, "Radio Address to the Nation on Administration Goals," radio address, January 31, 1987, at http://www.presidency.ucsb.edu/ws/index.php?pid=34674.

[77] NSF received $1.717 billion in appropriations in FY1988. P.L. 100-570 authorized NSF funding increases from FY1989 ($2.050 billion) through FY1993 ($3.505 billion). Actual appropriations to NSF in FY1993 were $2.734 billion, or $1.017 million (59%) more than the FY1988 funding level.

[78] Jeffrey Mervis, "Bush Signs NSF 'Doubling' Bill," *Science*, December 20, 2002, at http://news.sciencemag.org/ 2002/12/bush-signs-nsf-doubling-bill.

[79] NSF received $4.823 billion in appropriations in FY2002. P.L. 107-368 authorized NSF funding increases from FY2003 ($5.536 billion) to FY2007 ($9.839 billion). Actual appropriations to NSF in FY2007 were $5.890 billion, or $1.067 million (22%) more than the FY2002 funding level.

[80] Executive Office of the President, Domestic Policy Council, Office of Science and Technology Policy, *American Competitiveness Initiative: Leading the World in Innovation*, February 2006, at http://georgewbushwhitehouse.archives.gov/stateoftheunion/2006/aci /aci06-booklet.pdf.

[81] Other periods of time or funding units might produce different results. The decade-long perspective is largely consistent with the 11-year doubling pace enacted in the America COMPETES Reauthorization Act of 2010 (P.L. 111- 358).

[82] This growth estimate excludes American Recovery and Reinvestment Act (ARRA, P.L. 111-5) funding.

[83] This case is laid out more fully in National Academy of Sciences, National Academy of Engineering, and Institute of Medicine, Committee on Prospering in the Global Economy of the 21st Century: An Agenda for America Science and Technology, and Committee on Science, Engineering, and Public Policy, *Rising Above the Gathering Storm: Energizing and Employing America for a Brighter Economic Future*, National Academies Press, 2007, http://www.nap.edu/catalog/11463.html.

[84] Testimony of Boston University Associate Professor of Strategy and Innovation and National Bureau of Economic Research, Research Associate Dr. Jeffrey L. Furman, in U.S. Congress, Senate Committee on Commerce, Science, and Transportation, *Five Years of the America COMPETES Act: Progress, Challenges, and Next Steps*, hearings, 112th Cong., 2nd sess., September 19, 2012, at http://www.commerce.senate.gov/public/?a=Files.Serve& File_id=8687a045- afdc-4b74-ac75-efca96893a88.

[85] For example, a 2012 *Science* magazine report noted that the U.S. astronomical community had to revisit the priorities laid out in the 2010 Astronomy and Astrophysics Decadal Survey, which was drafted under the assumption of an NSF doubling, after it became clear that actual appropriations were not keeping pace with COMPETES act-authorized funding levels. See Yudhijit Bhattacharjee, "Panel Says NSF Should Shutter Six U.S. Instruments," *Science*, vol. 337, August 24, 2012, http://www.sciencemag.org/content /337/6097 /899. summary. Additionally, a September 2013 *Nature* editorial asserts that, at least in part due to the signal policy makers sent with the 2007 COMPETES act, NSF committed to two large ocean science division construction projects. Once completed, those projects will require operational support. *Nature* asserts that these operating costs will increase budget pressure on NSF's ocean science research account and that NSF should have anticipated that "big budgets would not last." See "Counting the Cost," *Nature* editorial, September 25, 2013, at http://www.nature.com/news /counting-the-cost-1.13804.

[86] Jules Duga and Tim Studt. "Government Spending Continues to Drive R&D Growth." *R&D,* vol. 47, no. 1 (2005), pp. F3-F7,F10-F15.

[87] Daniel Sarewitz, "Double Trouble? To Throw Cash at Science Is a Mistake," *Nature*, vol. 468, no. 135 (November 10, 2010), http://www.nature.com/news/2010/101110/full/468135a. html; and Daniel Sarewitz, "Blue-Sky Bias Should Be Brought down to Earth," *Nature*, vol. 481, no. 7339 (January 4, 2012), at http://www.nature.com/news/blue-skybias-should-be-brought-down-to-earth-1.9722.

[88] For more information about the federal STEM education effort, see CRS Report R42642, *Science, Technology, Engineering, and Mathematics (STEM) Education: A Primer*, by Heather B. Gonzalez and Jeffrey J. Kuenzi.

[89] For more information about STEM education funding at NSF, see CRS Report R42470, *An Analysis of STEM Education Funding at the NSF: Trends and Policy Discussion*, by Heather B. Gonzalez.

[90] Most NSF STEM education programs are operated under general authority.

[91] National Science Foundation, Graduate Research Fellowship Program, "About the NSG Graduate Research Fellowship Program," National Science Foundation Website, accessed on February 20, 2014, at http://www.nsfgrfp.org/about_the_program. NSF's *FY2015 Budget Request to Congress* seeks to increase the GRF stipend to $34,000. (Available at http://www.nsf.gov/about/budget/fy2015/index.jsp.)

[92] National Science Foundation, Integrative Graduate Education and Research Traineeship, "Introduction to the IGERT Program," *National Science Foundation Website*, accessed on February 20, 2014, at http://www.nsf.gov/crssprgm/igert/ intro.jsp.

[93] The GRF program total includes fellows across multiple cohorts. On average, the GRF program issues about 2,000 *new* fellowships each year. E-mail communication between CRS and staff from the NSF Office of Legislative and Public Affairs, dated February 21, 2014.

[94] Different legislators rejected the proposal—which included changes across a wide variety of programs and agencies—for different reasons. The "Joint Explanatory Statement" published in the January 15, 2014, *Congressional Record*, which accompanied P.L. 113-76 (Consolidated Appropriations Act, 2014), asserts that "the proposal contained no clearly defined implementation plan, had no buy-in from the education community, and failed to sufficiently recognize or support a number of proven, successful programs."

[95] For more information about the proposed FY2015 STEM education reorganization, see CRS Report IF00013, *The President's FY2015 Budget and STEM Education (In Focus)*, by

Heather B. Gonzalez; and CRS Report IN10011, *The Administration's Proposed STEM Education Reorganization: Where Are We Now?*, by Heather B. Gonzalez.

[96] Among other things, Section 101 of P.L. 111-358 directs the NSTC to develop a strategy for federal STEM education programs. NSTC published this strategy in June 2013. See Executive Office of the President, National Science and Technology Council, Committee on STEM Education, *Federal Science, Technology, Engineering, and Mathematics (STEM) Education: 5-Year Strategic Plan*, May 2013, available at http://www.whitehouse.gov/sites /default/files/ microsites/ostp/stem_stratplan_2013.pdf.

[97] H.Rept. 113-171, p. 59.

[98] House Committee on Science, Space, and Technology, Ranking Member Eddie Bernice Johnson, "Committee Discusses Proposed Reorganization of STEM Education Programs," press release, June 4, 2013.

[99] A webcast of the mark-up, as well as related hearing materials (including amendments and vote tallies), is available at http://science.house.gov/markup/subcommittee-research-and-technology-markup-hr-4186.

[100] NSF publishes its annual budget requests to Congress—dating from FY1998 to FY2015—on its website at http://www.nsf.gov/about/budget/. Additionally, policy makers may access award summaries by state and institution, historical NSF account data, and related reports at http://dellweb.bfa.nsf.gov/.

[101] NSF changed its account structure in FY2003. Prior years are not comparable. In FY2008, NSF shifted the EPSCoR program from EHR to RRA. Table B-2 treats EPSCoR as an RRA sub-account for all years in the data set.

[102] With one exception—the Expeditions in Education program, which was not included in the FY2014 request—these were the same programs included in the "OneNSF Framework" from NSF's FY2013 budget request.

[103] National Science Foundation, *FY2014 Budget Request to Congress*, April 10, 2013, http://www.nsf.gov/about/ budget/fy2014/index.jsp.

[104] Information about this initiative is included in the President's FY2015 budget request, available at http://www.whitehouse.gov/sites/default/files/omb/budget/fy2015/assets/budget.pdf.

[105] For more information about the NSF FY2014 budget request and appropriations, see CRS Report R43080, *Commerce, Justice, Science, and Related Agencies: FY2014 Appropriations*, coordinated by Nathan James, Jennifer D. Williams, and John F. Sargent Jr.

[106] With one exception—the Expeditions in Education program, which was not included in the FY2014 request—these were the same programs included in the "OneNSF Framework" from NSF's FY2013 budget request.

[107] National Science Foundation, *FY2014 Budget Request to Congress*, April 10, 2013, http://www.nsf.gov/about/ budget/fy2014/index.jsp.

[108] H.Rept. 113-171, S.Rept. 113-78, and the Joint Explanatory Statement published in the January 15, 2014, *Congressional Record*.

[109] For more information about the NSF FY2013 budget request and appropriations, see CRS Report R42440, *Commerce, Justice, Science, and Related Agencies: FY2013 Appropriations*, coordinated by Nathan James, Jennifer D. Williams, and John F. Sargent Jr.

[110] National Science Foundation, *FY2013 Budget Request to Congress*, February 13, 2012, p. Overview-3, http://www.nsf.gov/about/budget/fy2013/index.jsp.

In: National Science Foundation
Editor: Leslie Murphy
ISBN: 978-1-63321-972-4
© 2014 Nova Science Publishers, Inc.

Chapter 2

AN ANALYSIS OF STEM EDUCATION FUNDING AT THE NSF: TRENDS AND POLICY DISCUSSION*

Heather B. Gonzalez

SUMMARY

Federal policy makers have a long-standing interest in science, technology, engineering, and mathematics (STEM) education that dates to at least the 1st Congress. This interest is largely driven by concerns about the national science and engineering workforce, which is widely believed to play a central role in U.S. global economic competitiveness and national security.

The National Science Foundation (NSF) is a key component of the federal STEM education effort. Several inventories of the federal STEM education portfolio have highlighted NSF's important role—both in terms of funding and in the number and breadth of NSF programs. The NSF is also the only federal agency whose primary mission includes supporting education across all fields of science and engineering. As such, funding for STEM education at the NSF impacts not only the agency, but also the entire federal STEM education effort.

Congress reduced enacted funding levels (from the prior year) for NSF's main education account in both FY2011 and FY2012. Those year-

* This is an edited, reformatted and augmented version of a Congressional Research Service publication, No. R42470, dated January 31, 2014.

over-year reductions followed several years of varying funding, as well as changes in the overall distribution of the foundation budget that reduced funding for the main education account as a percentage of the total NSF budget. For the most part, these changes appear to result from a combination of holding the main education account more or less constant while applying most of the foundation's FY2003-FY2012 budget growth to the main research account. However, in constant dollar terms, it appears at least some of the increase in funding for research activities during the observed period may have come at the expense of education activities.

It is not clear if these funding changes reflect evolving congressional and Administration policy priorities and an intentional prioritization of research over educational activities at the NSF, or if they reflect the cumulative impact of funding decisions made in response to specific conditions in specific fiscal years that happen to have had this effect. Further, the significance of these changes for NSF's STEM education and research missions—and for the overall federal STEM effort— depends, in part, on how they fit within the broader policy context. In particular, it depends (among other things) on how policy makers perceive and assess the policy rationale behind STEM education funding at the NSF; the character of NSF's STEM education activities; the foundation's role in the federal STEM education portfolio; and the impact of changes in NSF's education account on the foundation's other primary mission, research.

This report analyzes NSF funding trends and selected closely related STEM education policy issues in order to place conversations about NSF's budget in a broader fiscal and policy context. It concludes with an analysis of potential policy options.

INTRODUCTION

Federal policy makers have a long-standing interest in science, technology, engineering, and mathematics (STEM) education. This interest is largely driven by concerns about the national science and engineering workforce, which is widely believed to play a central role in U.S. global economic competitiveness and national security. The U.S. STEM education system is a primary source of scientists and engineers in the United States.[1] Approximately 6% of the U.S. workforce was employed in a STEM field in 2011. About 70% of those workers had at least a bachelor's degree. Further, of college graduates who were employed in a STEM field, 73% had a science or engineering major.[2]

Given the oft-cited connection between STEM education and key national priorities, federal policy makers have historically paid close attention to the U.S. STEM education system. The federal STEM education effort is wide-ranging. Analysts have identified between 105 and 252 STEM education programs and activities at 13 to 15 federal agencies. Annual federal appropriations for STEM education are typically in the range of $2.8 billion to $3.4 billion. Published inventories of the federal STEM education effort identify the Department of Education (ED), National Science Foundation (NSF), and Department of Health and Human Services (HHS) as key agencies in the federal effort. Of these, the NSF has the most STEM education funding and largest number of programs (typically). The foundation is also the only federal agency whose primary mission includes supporting education across all fields of science and engineering. As such, the NSF is a key component of the federal STEM education portfolio.[3]

Funding for STEM education decreased as a percentage of the total NSF budget between FY2003 and FY2012.[4] Further, funding levels for the foundation's main education account were lower than the previous year in both FY2011 and FY2012. FY2013 current plan funding for the main education account appears to be close to FY2012 levels.[5] The significance of these funding trends for NSF's education and research missions, as well as for the federal STEM education effort overall, depends in part on how these changes fit within historical funding trends at the NSF. This report analyzes those trends—and addresses selected STEM education policy issues—in order to place the conversation about federal funding for STEM education at NSF in broader fiscal and policy context.

Methodology, Sources, Data, and Notes

This report examines actual funding for the NSF from FY2003 to FY2012 in current and constant (2005) dollars. Congress provides appropriations in current dollars, so current dollar funding data align with annual appropriations measures and congressional actions, while constant dollar data adjust for the effects of inflation and provide insight into purchasing power. This report also analyzes the distribution of total NSF funding by appropriations account and by character class. The character class analysis adjusts for programs that draw from more than one appropriations account but serve the same program or activity. Over time, changes in the distribution of funding may reflect changing policy priorities.

Several other introductory points should be noted.

- This report uses the following terms for the major appropriations accounts at NSF: "R&RA" or "main research account" for Research and Related Activities, "E&HR" or "main education account" for Education and Human Resources, "MREFC" or "main construction account" for Major Research Equipment and Facilities Construction, "AOAM" for Agency Operations and Awards Management, "NSB" for National Science Board, and "OIG" for Office of the Inspector General.
- This report uses the following terms for major activities: "R&D" or "research activities" for research and development-related activities, "E&T" or "education activities" for education and training-related activities, and "NIA" for non-investment activities. Non-investment activities are primarily administrative activities (e.g., travel and compensation costs for proposal review panelists).
- The analysis in this report is based on budgetary data from the NSF's annual budget requests to Congress from FY2005 to FY2014 and from information provided to CRS by NSF. Appropriations account data come from the "Overview" sections of the NSF budget requests; funding data by character class come from the "Quantitative Data Table" sections.
- CRS adjusted the appropriations account data for FY2003 to FY2005 to reflect the transfer of the Experimental Program to Stimulate Competitive Research (EPSCoR) between major accounts. This analysis treats EPSCoR as a research account program for all years in the data set.
- NSF adopted its current appropriations account structure in 2003. FY2003- FY2012 are the most recent years for which actual funding data are available and comparable.
- NSF programs are often co-funded (e.g., funded by two or more appropriations accounts). Budgetary data that are broken down by character class adjust for co-funding and provide insight into what NSF actually spends on a given activity.
- Funding levels for FY2009 and FY2010 do not include funding from the American Recovery and Reinvestment Act (AR&RA, P.L. 111-5) because NSF treated these funds as supplemental in its budget calculations.

- To generate constant dollar (e.g., inflation-adjusted) data in FY2005 dollars CRS used the Office of Management and Budget (OMB) deflator published in Table 10.1 of the OMB's *Historical Tables* and accessed on December 23, 2013.
- Data used in this report may be found in **Appendix A** and **Appendix B**.

HISTORICAL FUNDING TRENDS AT NSF

By Character Class

The NSF's budget can be broken down by character class.[6] According to the NSF, its two primary activities are research and development (R&D) and education and training (E&T). The NSF also has a category for what it calls "non-investment activities," or NIA, which pays for items such as proposal review panel travel and compensation costs, invitational travel, and other administrative activities.[7] Unlike appropriations accounts, which show how Congress provides funding to the foundation, the character class perspective adjusts for co-funding (i.e., when programs are supported by more than one appropriations account) and provides insight into total funding for the foundation's two primary missions (e.g., research and STEM education).

Figure 1 shows how NSF current dollar funding for E&T and R&D activities changed between FY2003 and FY2012. (See **Appendix A** for data.) The total current dollar increase in NSF funding between FY2003 and FY2012 was $1.735 billion, or 32% more than the FY2003 baseline of $5.369 billion. During this period, current dollar funding for R&D increased by $1.762 billion while current dollar funding for E&T decreased by $95 million. Funding for R&D increased more or less steadily, while funding for E&T fluctuated. E&T funding levels ranged from a high of $941 million in FY2004 to a low of $783 million in FY2006. The median funding level for E&T during the observed period was $840 million. FY2012 funding for E&T was below the FY2003 level. When expressed in constant (2005) dollars, funding for R&D experienced a relatively flat period from FY2005 through FY2008. (See **Figure 2**.) The sharp increase in constant dollar R&D funding between FY2009 and FY2010 coincides with the second year of funding under the America COMPETES Act (P.L. 110-69).[8] R&D funding levels have stayed about the same since then. Constant dollar funding for E&T generally trended downward between FY2003 and FY2012—ranging from a high of $972 million in FY2004 to a low of $696 million in FY2012. Median constant dollar funding for E&T activities during the observed period was $797 million.

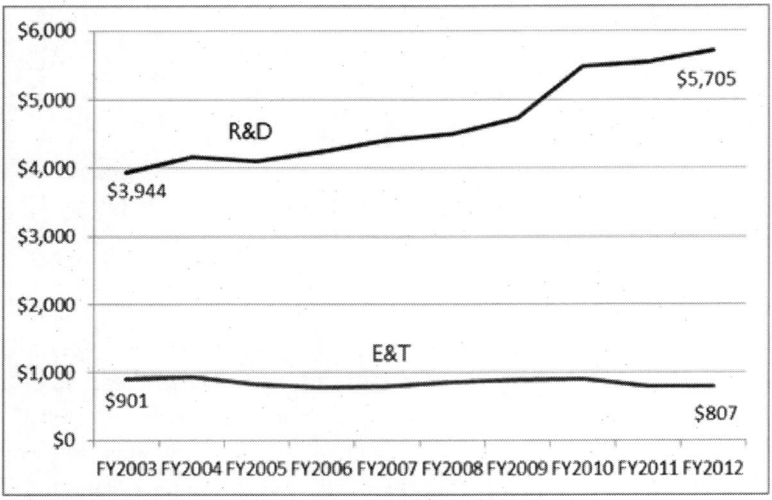

Source: CRS analysis based on data provided in annual NSF budget requests to Congress (FY2005 to FY2014).

Figure 1. NSF Funding by Character Class (Current); FY2003 to FY2012 Actual, In Millions, Rounded.

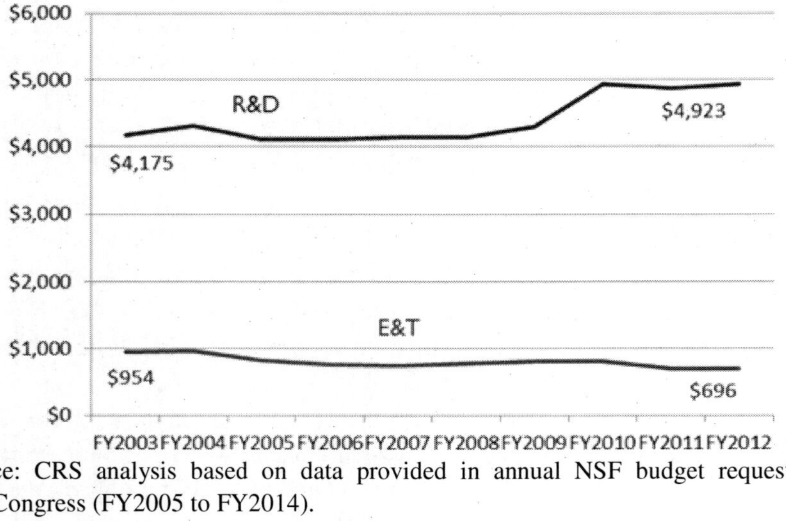

Source: CRS analysis based on data provided in annual NSF budget requests to Congress (FY2005 to FY2014).

Figure 2. NSF Funding by Character Class (2005 Constant); FY2003 to FY2012 Actual, In Millions, Rounded.

In constant (2005) dollars, total NSF funding increased by $447 million—from $5.684 billion to $6.131 billion—between FY2003 and FY2012. Funding for R&D increased by $748 million while funding for E&T decreased by $258 million during this same period. (See **Figure 3**.) Given that funding for R&D grew in excess of the total NSF increase, and that funding for E&T experienced negative growth, at least some of the growth in funding for R&D between FY2003 and FY2012 appears to have come from E&T.

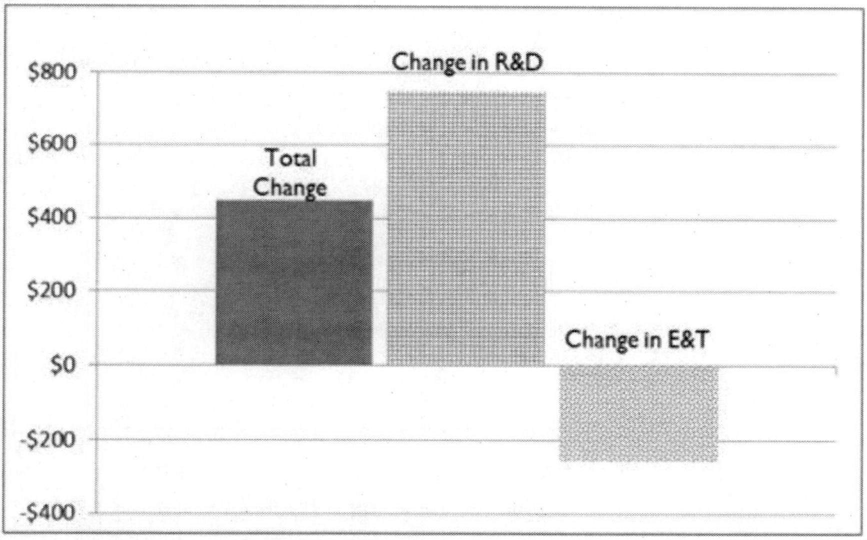

Source: CRS analysis based on data provided in annual NSF budget requests to Congress (FY2005 to FY2014).

Figure 3. Change in NSF Funding from FY2003 to FY2012, by Character Class; Total Change, Change in R&D, and Change in E&T In Millions of Constant (2005) Dollars, Rounded.

The distribution of NSF funding by character class also changed between FY2003 and FY2012. As **Table 1** shows, the percentage of the NSF budget dedicated to E&T activities generally decreased while the percentage dedicated to R&D activities generally increased. This is consistent with the previous finding that most of the total increase in NSF funding during the observed period went to R&D. Some of the increase in the percentage of the NSF budget dedicated to R&D may have come from NIA. (See "Note," **Table 1**.)

Table 1. Distribution of NSF Obligations by Character Class (% of Total) FY2003 to FY2012 Actual and FY2003 to FY2012 Average

Class	FY 2003	FY 2004	FY 2005	FY 2006	FY 2007	FY 2008	FY 2009	FY 2010	FY 2011	FY 2012	Ave. '03-'12
R&D	73.5%	73.7%	74.8%	75.1%	74.9%	74.1%	73.1%	78.7%	80.1%	80.3%	75.8%
NIA	9.8%	9.6%	10.2%	11.0%	11.5%	11.9%	13.0%	8.4%	8.3%	8.3%	10.2%
E&T	16.8%	16.7%	15.0%	13.9%	13.6%	14.1%	13.9%	12.9%	11.6%	11.4%	14.0%

Source: CRS analysis based on data provided in annual NSF budget requests to Congress (FY2005 to FY2014).

Note: In response to direction from the Office of Management and Budget, NSF reclassified certain NIA obligations as R&D in FY2010. It is likely that pre-FY2010 R&D levels are higher than represented here, but the amount of the difference is unknown. NSF indicates that there may be a coding problem in the FY2008 E&T numbers, such that some R&D activities may be improperly assigned to the E&T account.

By Appropriations Account

Congress provides funding to the NSF via appropriations accounts, not by character class. Analysis of appropriations account trends, therefore, aligns more closely with congressional action than analysis by character class. This section focuses on changes in congressional appropriations for the R&RA and E&HR accounts because these accounts are the primary sources of support for NSF mission activities. As with the previous analysis of character class trends, this section examines both current and constant (2005) trends in funding for these accounts.

Total current dollar appropriations to NSF increased from $5.369 billion to $7.105 billion between FY2003 and FY2012. Of this $1.735 billion increase, 93.0% or $1.615 billion went to R&RA. Current dollar appropriations for E&HR decreased by $15.2 million (-1.8%) during the same period. **Figure 4** shows the current dollar trends in R&RA and E&HR from FY2003 to FY2012. (See **Appendix B** for data.)

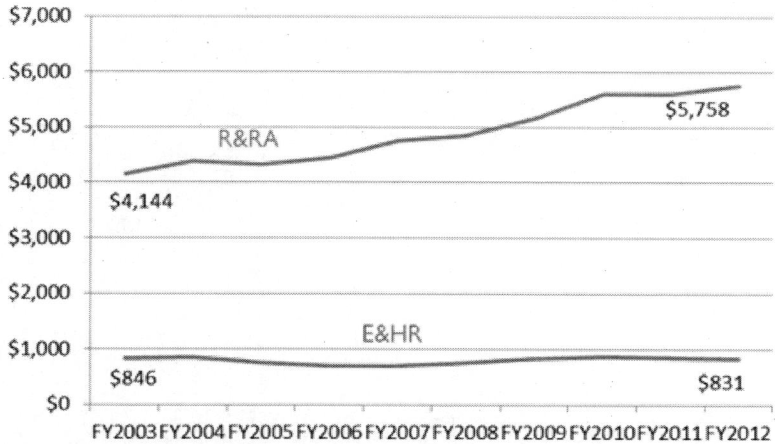

Source: CRS analysis based on data provided in annual NSF budget requests to Congress (FY2005 to FY2014).

Figure 4. NSF Funding for R&RA and E&HR (Current); FY2003 to FY2012 Actual, In Millions, Rounded.

Both R&RA and E&HR experienced current dollar reductions from the prior year in FY2005 and FY2011. However, other than in these years, R&RA funding levels increased over the prior year for each year in the observed period. E&HR funding varied. It received four year-over-year increases and five year-over-year reductions between FY2003 and FY2012. Further, reductions to the E&HR account appeared to be steeper, and took longer to return to pre-reduction levels, than did reductions to R&RA.[9]

In constant (2005) dollars, NSF funding increased by $447 million between FY2003 and FY2012. As shown in **Figure 5**, constant dollar funding for R&RA increased by $583 million (13%) while constant dollar funding for E&HR decreased by $179 million (-20%). These trends suggest that most of the total constant dollar growth at the NSF—and at least some of the constant dollar value of E&HR—accrued to R&RA during the observed period.

As **Table 2** shows, the percentage of the NSF budget dedicated to R&RA activities has generally increased and the percentage of the budget dedicated to E&HR has generally decreased since FY2003.

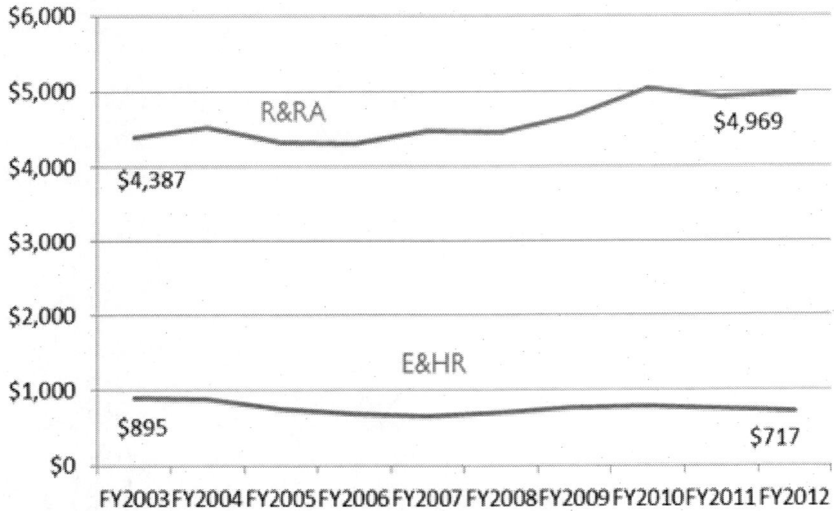

Source: CRS analysis based on data provided in annual NSF budget requests to Congress (FY2005 to FY2014).

Figure 5. NSF Funding for R&RA and E&HR (2005 Constant); FY2003 to FY2012 Actual, In Millions, Rounded.

Table 2. Distribution of NSF Funding by Appropriations Account (% of Total) FY2003 to FY2012 Actual and FY2003 to FY2012 Average

Account	FY 2003	FY 2004	FY 2005	FY 2006	FY 2007	FY 2008	FY 2009	FY 2010	FY 2011	FY 2012	Ave. '03-'12
R&RA	77.2%	77.6%	79.0%	78.8%	80.9%	79.8%	79.7%	80.5%	81.1%	81.0%	79.6%
E&HR	15.7%	15.0%	13.7%	12.4%	11.8%	12.6%	13.1%	12.5%	12.5%	11.7%	13.1%
MREFC	3.3%	3.3%	3.0%	4.1%	2.8%	2.7%	2.5%	2.4%	1.8%	2.8%	2.9%
AOAM	3.5%	3.9%	4.1%	4.4%	4.2%	4.6%	4.5%	4.3%	4.3%	4.2%	4.2%
NSB	0.1%	0.0%	0.1%	0.1%	0.1%	0.1%	0.1%	0.1%	0.1%	0.1%	0.1%
OIG	0.2%	0.2%	0.2%	0.2%	0.2%	0.2%	0.2%	0.2%	0.2%	0.2%	0.2%

Source: CRS analysis based on data provided in annual NSF budget requests to Congress (FY2005 to FY2014).

Policy Issues and Observations

As the previous section's analysis of historical funding trends at NSF shows, Congress reduced funding for NSF's main education account in both FY2011 and FY2012. Those year-over-year reductions followed several years of fluctuating funding for E&HR. In addition, changes in the distribution of the foundation budget reduced funding for the main education account as a percentage of the total NSF budget. These changes generally appear to result from a combination of holding the main education account more or less constant while applying most of the foundation's FY2003-FY2012 budget growth to R&RA. However, in constant (2005) dollars, at least some of the growth in NSF research funding appears to come from the foundation's education-related activities.

It is not clear if these funding changes reflect evolving congressional and Administration policy priorities and an intentional prioritization of research over educational activities at the NSF. They may simply reflect the cumulative impact of funding decisions made in response to specific conditions in specific fiscal years. Further, methodological and research limitations can impede the types of economic analyses that might otherwise be used to assess optimal funding levels for NSF accounts. The period of time for which comparable budgetary account data exist, for example, is limited to 10 years. It can also be difficult to definitively link federal investments in research and education *writ large* with social, scientific, or economic outcomes.

However, historical funding trends at the NSF raise several questions for Congress as it considers funding for the foundation, as well as for the federal STEM education effort overall. These questions include

- What is the policy rationale behind funding for STEM education at NSF?
- What are NSF's STEM education activities?
- What is NSF's role within the federal STEM education portfolio?
- What impact might changes in the NSF STEM education account have on research activities at NSF?
- What are the policy options for Congress as it considers future NSF budgets?

What Policy Rationale Drives Funding for STEM Education at NSF?

One of the main reasons that advocates support increased funding for STEM education programs at NSF is because of their perceived contribution to the U.S. science and engineering (S&E) workforce. A broad consensus of business, academic, and policy leaders holds that U.S. STEM education weaknesses have or will soon contribute to national S&E workforce shortages and that this labor supply problem has or will diminish U.S. global economic competitiveness and threaten national security.[10] Analysts who hold this view typically contend that the federal government should increase funding across, or within specific parts of, the so-called STEM education "pipeline" (pre-kindergarten to post-graduate education). These investments, advocates assert, will improve U.S. student performance in STEM subjects and increase both the quantity and quality of U.S. students graduating with degrees in STEM fields. Programs designed to improve teaching and learning in STEM fields or to attract and retain students in STEM degree programs through scholarships and financial aid are examples of this policy approach.

Other observers counter that U.S. students are not underperforming in mathematics and science; that shortage claims are overstated, misunderstood, or do not call for supply side interventions;[11] and that demand, not supply, may be the bigger policy challenge.[12] These analysts assert that the United States graduates more science and engineering students than there are science and engineering jobs[13] and that classic signs of labor shortages (e.g., rapidly increasing wages) are not broadly evident in the U.S. STEM labor supply. These analysts typically acknowledge that there may be reasons to seek improved student STEM performance, but they argue that the current policy debate is based on misperceptions, obscures root causes of poor performance (e.g., poverty effects in education), and results in ineffective policy responses.[14] Analysts who hold this view suggest demand-side policies (e.g., increased funding for R&D, tax credits for privately funded research, or more and better jobs for scientists), improved labor market signaling, and addressing the root problems of low-performing students (e.g., poverty) as possible policy alternatives.

A third view of the shortage issue asserts that the disagreement about the adequacy of the supply of STEM workers "can be resolved by the fact that large numbers of people with STEM talent or degrees divert from STEM occupations either in school or later in their careers."[15] Analysts who hold this view contend that the economy increasingly values and demands STEM competencies— for example, the knowledge, skills, and abilities typically associated with education in STEM fields—even in non-STEM occupations. These analysts assert that workforce shortages become more evident if demand from *both* STEM *and* non-STEM fields is compared to the supply of STEM-educated workers. (In other words, they assert that analysts who argue that there are no shortages are missing part of the equation.) Those who hold this view typically agree that a supply response is appropriate, and recommend paying more attention to the role of personal interest in career choices, nurturing students with a personal interest in STEM "even if they do not look like traditional STEM workers,"[16] and integrating STEM competencies into a broader array of academic disciplines.

What Are NSF's STEM Education Activities?

Funding for STEM education at NSF serves a variety of objectives. However, a 2011 report found that 66% of NSF's STEM education budget provided for programs designed either to support postsecondary students (primarily through scholarships and other forms of financial support) or for research on teaching and learning in STEM fields. Most of the postsecondary student funding went to the Graduate Research Fellowship (GRF) and Integrative Graduate Education and Research Traineeship (IGERT) programs, which provide stipends and support to STEM graduate students. Most of the funding for research in STEM education went to the Discovery Research K12 and Mathematics and Science Partnership programs, which seek to improve kindergartenthrough-Grade 12 (K-12) STEM education. Smaller portions of NSF's STEM education budget provided for a number of other objectives.[17] (See **Figure 6**.)

Several reports on the federal STEM education effort have noted a general dearth of STEM education program evaluations and have recommended that federal agencies increase their program evaluation rates. This challenge is not broadly applicable to the NSF, which has conducted evaluations of many of its STEM education programs. However, in a January 2012 Government Accountability Office (GAO) review of federal STEM education programs, GAO found that federal STEM education program evaluations—including NSF evaluations—could be improved. In particular, GAO recommended improved survey response rates, better alignment of methods with other components of the evaluation, and robust use of criteria to measure outcomes.[18]

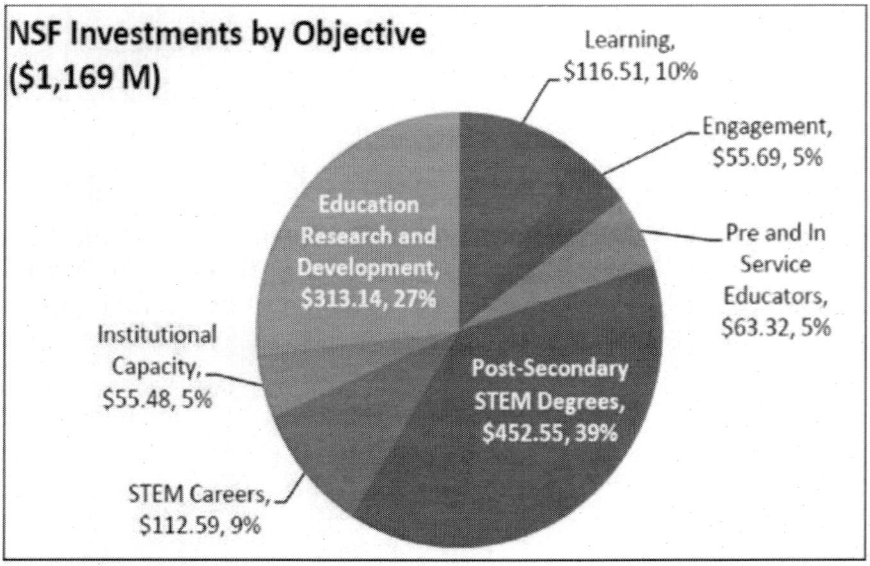

Source: Executive Office of the President, National Science and Technology Council, Committee on STEM Education, Fast-Track Action Committee on Federal Investments in STEM Education, *The Federal Science, Technology, Engineering, and Mathematics (STEM) Education Portfolio*, December 2011, p. 68.

Notes: The NSTC calculation for STEM education funding at the NSF in FY2010 (actual) uses a different methodology than either the E&HR or E&T estimates used in other parts of this report.

Figure 6. STEM Education Funding at NSF, by Objective. FY2010 actual, millions of current dollars.

What Is NSF's Role in the Federal STEM Education Portfolio?

The NSF plays a key role in the federal STEM education portfolio. For example, the National Science and Technology Council (NSTC) estimates that total federal STEM education investments were $3.4 billion in FY2010.[19] The NSF portion of that total was $1.2 billion (rounded). The NSF is also the only federal agency whose primary mission includes education across all fields of science and engineering. This key position means changes at the NSF may disproportionally affect the entire federal STEM education effort (both funding and character).

In terms of the character of its contribution to the federal STEM education portfolio, NSF highlights its STEM education research and development (R&D) functions. The foundation states that it focuses on identifying effective

STEM education practices through research and small-scale testing, but that it is not well-positioned to bring these practices to scale.[20]

NSF is also an important source of scholarships, fellowships, and financial support to STEM students as well as institutions of higher education. For example, since the establishment of the Graduate Research Fellowship (GRF) program in 1952—two years after NSF's own founding in 1950—NSF has supported researchers and students in STEM fields. This funding serves integrated research and education purposes. It seeks to support the national research effort through support of the STEM workforce and it seeks to support the national STEM education effort by providing financial and educational incentives for students to go into STEM and STEM-related fields (such as K-12 science teaching). NSF estimates that it provides financial support to about 5% of the science and engineering graduate students in the United States.[21]

In addition to NSF's role as a funder of STEM education R&D and STEM student support, the foundation also operates smaller (measured by funding levels) programs that seek to advance other federal STEM education policy priorities. These include programs designed to increase the participation of historically under-represented groups in STEM fields and programs that provide funding for out-of-school or informal STEM education. Some NSF STEM education programs are also integrated with similar programs at other agencies, such as the Mathematics and Science Partnership program, which has a sister program at ED.

Although NSF is a major contributor to the federal STEM education portfolio, some analysts may argue that national STEM education objectives could be met without some (or all) NSF STEM education programs.

The Federal STEM Education Portfolio: Selected Governance Concerns

The specific debate about funding for STEM education programs at the NSF is taking place within a broader conversation about governance of the federal STEM education portfolio. This conversation has focused on the potential for duplication in the federal effort and on the perception that the federal effort lacks both coordination and an overarching strategy. The Obama Administration proposed a reorganization of the federal STEM education portfolio as part of the FY2014 budget request. The proposed reorganization envisions a prominent role for NSF in the federal STEM education portfolio. Policy makers have also expressed ongoing concern about the dissemination of NSF-funded STEM education research. These issues are discussed in greater detail in the following sections.

Duplication and Consolidation

The scope, scale, and perceived lack of coordination in the federal STEM education portfolio have some analysts concerned that federal agencies are duplicating effort. In response to these concerns, some policy makers have proposed consolidating or eliminating some or all of NSF's STEM education programs.[22]

Published assessments of duplication in the federal STEM education portfolio are somewhat contradictory. Preliminary findings from an April 2011 GAO report appeared to suggest the potential for duplication in federal teacher quality programs, including teacher quality programs at the NSF.[23] However, the December 2011 NSTC comprehensive inventory of federal STEM education programs specifically examined the duplication question within the federal STEM education portfolio and found "little overlap and no duplication."[24] A January 2012 GAO report on the federal STEM education effort concluded that 83% of federal STEM education programs overlapped "to some degree," but stated that this overlap would "not necessarily be duplicative."[25]

Federal program consolidation is a widely debated option that policy makers may employ to reduce duplication and potentially affect savings. Some policy makers see program consolidation as a means to increase program flexibility and improve program responsiveness, because federal program managers would have greater authority to shift priorities without having to modify federal law.[26] However, other policy makers may object to this change, because it can transfer program control from the legislative to the executive branch, potentially shifting the balance of power between the branches. Consolidation (particularly in the form of block grants) has also been proposed as a strategy to transfer control to the states and as a means to reduce program costs. Such a shift could increase the ability of states to respond to local conditions and needs, but might make it more difficult for federal policy makers to implement a national STEM education agenda,[27] or to leverage the unique assets that federal science agencies bring to the STEM education effort.[28] On the issue of cost and consolidation, the GAO has found that program consolidation can be more expensive in the short term and may not result in long-term savings if program workloads are not reduced.[29] Consolidation opponents raise general concerns about the potential impact of merging programs, arguing that certain programs (such as STEM education programs) need specified funding streams to avoid being passed over in favor of competing educational priorities.

The impact of federal STEM education program consolidation efforts on STEM education at the NSF will depend on what programs are consolidated, how the consolidation is accomplished, how funding streams are affected, and the degree to which NSF programs are strictly duplicative of other federal STEM education efforts.[30] Congress could, for example, seek either a full or partial consolidation of STEM education programs at either the NSF or across the entire federal STEM education portfolio. Savings and program impacts would vary, depending on which of these strategies policy makers pursue.

A Federal STEM Education Strategy

A second policy issue raised in the current federal STEM education governance debate relates to the perceived lack of coordination or an overarching strategy in the portfolio. Until recently, the federal STEM education effort was largely unknown and primarily undertaken in a distributed fashion that responded to the specific needs of agencies and STEM constituencies. Programs were typically not part of a defined, overarching federal STEM education strategy or well-coordinated approach across federal agencies. Although some analysts may view the distributed method as particularly responsive to the unique workforce needs or STEM education assets of federal science agencies, other observers have suggested that an overarching cross-agency strategy may improve the efficiency of federal STEM education investments.[31]

Both Congress and the Administration have moved to develop a federal STEM education strategy. Section 101 of the America COMPETES Reauthorization Act of 2010 (P.L. 111-358) directed the NSTC to develop and implement a five-year federal STEM education strategy. (NSF co-chairs the NTSC subcommittee on STEM education, colloquially known as "CoSTEM," that was formed in response to Section 101.) The NSTC issued a report on the status of the strategy in February 2012 (hereinafter referred to as the "status report").[32] The status report identified two STEM education goals—STEM workforce development and STEM literacy—as well as policy and administrative strategies designed to accomplish these goals. Further, the status report identified four priority policy areas for the federal effort: "effective K-12 teacher education, engagement, undergraduate STEM education, and serving groups traditionally underrepresented in STEM fields."[33] Noting that strong arguments can be made for other STEM education policy areas, the report states that these priority areas were chosen because they represent the convergence of "national needs, Presidential priorities, and federal assets."[34] Agencies would retain the authority to establish their own STEM education priorities as well.

The NSTC published its five-year strategic plan for federal STEM education investments (hereinafter referred to as the "strategic plan") in May 2013. The strategic plan envisions a future where

> The United States has a well-qualified and increasingly diverse STEM workforce able to lead innovation in STEM-related industries and to fulfill CoSTEM agency workforce needs; American students have access to excellent P-12, postsecondary, and informal STEM education and learning opportunities; and Federal STEM education programs are based on evidence and coordinated for maximum impact in priority areas.[35]

To achieve this vision, the strategic plan identifies five priority areas for federal STEM education investment: improve STEM instruction, increase and sustain youth and public engagement in STEM, enhance STEM experiences of undergraduate students, better serve groups historically underrepresented in STEM fields, and design graduate education for tomorrow's STEM workforce. Within each of these priority areas, the strategic plan identifies specific goals, such as preparing 100,000 excellent new K-12 teachers by 2020 and supporting a 50% increase in the number of U.S. youth who have a STEM experience each year. The strategic plan includes an outline for implementation, but notes that "detailed roadmaps" would need to be developed.[36]

Congressional response to the strategic plan has, to some degree, become part of a broader conversation about the FY2014 Administration budget request. In April 2013, prior to the publication of the strategic plan, the Obama Administration released its FY2014 budget request. Integrated into the request was a proposal to reorganize the federal STEM education effort.[37] The Administration asserts that the proposed reorganization was informed by drafts of the strategic plan.[38] However, some policy makers assert that the strategic plan "appears to have been modified at the last minute to bring it into conformance with the Administration's STEM education budget proposal," and argue that in so doing, "key elements of the progress report were lost or diluted."[39] On the other hand, some policy makers who object to the proposed reorganization offered the strategic plan as "a new starting point for discussion" about any potential changes to the federal STEM education effort.[40]

The adoption and implementation of an overarching federal STEM education strategy could have many implications for STEM education at the NSF, depending on the type of strategy policy makers adopt and the STEM education goals they pursue. The America COMPETES Reauthorization Act of 2010 gave the executive branch the authority to both develop and implement a federal STEM education strategy. However, implementation depends on appropriations and related congressional decisions. If the 113th Congress adopts the NSTC strategy or proposed FY2014 reorganization in its appropriations and authorization actions, NSF will likely continue its large role in the national STEM education strategy. If legislators pursue similar goals, but undertake a different strategy—such as increasing funding for Advanced Placement course-taking or providing more funding for early childhood education at ED—then the NSF may play a different role. Alternatively, legislators may adopt different national STEM education goals or strategies.

Obama Administration Proposal to Reorganize the Federal STEM Education Effort in FY2014

The Obama Administration's FY2014 budget request includes a proposal to reduce the number of federal STEM education investments by about half while increasing total funding for federal STEM education activities by about 6% over FY2012 funding levels (hereinafter referred to as the "proposed reorganization"). Under the proposed reorganization, approximately 78 programs at nine federal agencies would lose funding for certain STEM education activities. Some STEM education activities within agencies would also be consolidated. Funding for certain priority programs at three "lead" agencies (NSF, ED, and the Smithsonian Institution) would increase. Under the proposal, NSF would focus on improving undergraduate STEM education as well as creating a more coherent system of federal graduate fellowships.

The Administration asserts that the proposed reorganization would "cut back lower-priority or narrow-purpose programs to make room for targeted increases."[41] The Administration further argues that the proposed reorganization would decrease fragmentation in the federal STEM education effort, "allowing potential for easier coordination and strong evaluations of what's working."[42] Some policy makers share the Administration's broad concerns about fragmentation—and perceive a similar lack of coordination—in the federal STEM education effort. Other policy makers also express support for consolidation, generally, as a response to perceived duplication in the federal STEM education effort. The proposed reorganization, which includes both a coordinated approach and a program consolidation, potentially addresses those concerns for some policy makers.

The proposed reorganization has received a mixed response from Congress—some policy makers accept certain proposed changes, while others seek to prevent implementation.[43] Objection to the proposed reorganization stems from a variety of factors, most of which relate to the process by which the proposal was developed (or perceived to have been developed) and its timing.

For example, some policy makers assert that the proposed reorganization was developed outside of, or with insufficient insight from, the process for coordinating federal STEM education programs that Congress established in Section 101 of the America COMPETES Reauthorization Act of 2010 (P.L. 111-358). In a June 2013 House Committee on Science, Space, and Technology hearing, several legislators expressed concern about the release of the proposed reorganization prior to the publication of the strategic plan for federal STEM education as mandated by Section 101.[44] Similarly, the Senate Appropriations Committee report on FY2014 funding for Commerce, Justice, Science, and Related Agencies

> defers action on the [proposed reorganization] until such time that [the Office of Science and Technology Policy] ... finalizes the STEM program assessments as required by America COMPETES.[45]

Other stakeholders argue that the Administration's process for developing the proposed reorganization was insufficiently transparent—the proposal is reported to have surprised many in the scientific community—and that too few stakeholders were consulted in its development. Further, although the Administration asserts that the proposed reorganization will facilitate better program evaluation, news reports state that that "previous evaluations of STEM programs were not the driving force in selecting winners and losers."[46] Additionally, some policy makers question the capacity of lead agencies to take on their new roles or express support for programs at ceding agencies, like the National Aeronautics and Space Administration.

The effect of the Administration's proposed reorganization of federal STEM education activities on STEM education programs at the NSF is unclear. A change of this magnitude—approximately half the federal effort—could result in a wide range of intended and unintended consequences. The Administration has not provided Congress with a detailed implementation plan or impact analysis. Further, many of the proposed changes require either tacit or specific approval from multiple congressional committees, as well as from Congress as a whole.

Dissemination

The dissemination of NSF's STEM education research—including research evaluating the effectiveness of NSF STEM education programs—to other federal agencies and education stakeholders is an ongoing policy challenge.

Some policy makers have responded to the dissemination challenge by seeking improved collaboration between federal agencies at both the portfolio and program levels. At the portfolio level, for example, the NSTC's federal STEM education strategy proposes sharing evidence-based approaches (e.g., established by research) as a primary strategy toward accomplishing federal STEM education goals.[47] At the program level, the Administration's FY2014 budget request seeks funding for at least two STEM education collaborations between NSF and ED.[48] Whether these collaborations (if funded) will prove successful depends on program managers' willingness to collaborate, on executive branch leadership support for collaboration, and on the institutional cultures of the respective agencies, among other things.

Other strategies to address the dissemination challenge include policies directing NSF to independently distribute STEM education research to stakeholders. For example, the House Appropriations Committee report on Commerce, Justice, Science, and Related Agencies Bill, 2012 (H.Rept. 112-169) directed NSF to independently distribute research on best practices in STEM education to stakeholders. H.Rept. 112-169 also directed the NSF to develop methods to track and evaluate stakeholders' implementation of that research and to report to Congress on progress.[49] In response to these and related congressional directives, NSF funded a National Research Council study on successful STEM education and held a series of workshops around the country on promising practices in STEM education.[50]

What Impact Might Changes in the NSF STEM Education Account Have on Research Activities at NSF?

Although this report focuses on NSF STEM education programs, many of those programs are co-funded (e.g., they receive funding from other foundation accounts, principally R&RA). As such, changes in the main NSF education account may impact the main research account. For example, the

R&RA contribution to E&T activities increased by 85% (from $143 million to $264 million) between FY2003 and FY2012. Some of this increase may be attributable to policy changes made by Congress and the President. Section 510 of the America COMPETES Reauthorization Act of 2010 (P.L. 111-358) requires NSF to fund the GRF program with equal contributions from both R&RA and E&HR. NSF has increased the R&RA contribution to the GRF program in accordance with Section 510. Further, because the GRF program is typically classified as an E&T activity, this change may account for some of the increase in R&RA funding for E&T activities. However, R&RA funding for E&T activities was increasing prior to the enactment of Section 510.

The impact of these funding dynamics on NSF's research capacity is unclear. FY2012 R&RA funding for E&T was $264 million, which was high compared to the historic average for E&T funding in the R&RA account, but was still less than 5% of the total R&RA budget that year ($5.758 billion). One interpretation of these changes in R&RA is that they represent a deeper integration of the foundation's complementary research and education missions. The research experience is sometimes perceived as serving educational purposes,[51] just as support for education may ultimately benefit the research enterprise. However, an alternative explanation might be that budgetary fluctuations or stagnation in the main education account put pressure on the main research account. Congress may wish to consider whether and to what extent the R&RA account should serve education and training purposes and whether the main education account is sufficient for congressional priorities in these areas.

OPTIONS FOR CONGRESS

Among the several options available to Congress are the following.

- **Maintain NSF budget as it is.** If Congress seeks to preserve NSF's budgetary autonomy, it could maintain the NSF budget as it is, making no significant increases or reductions and without directing the agency to change the distribution of funding between its main research and education accounts. This would provide the foundation its historical discretion, while Congress could continue to guide its activities through the oversight process.
- **Increase funding for STEM education at NSF.** If Congress seeks to increase NSF's STEM education capacity and its role in the federal

portfolio, it could provide additional funding for NSF's STEM education activities. It could

- *Increase funding for E&HR*—Congress could do this by either shifting funding from other appropriations accounts to E&HR or by providing additional funding directly to the main education account. A shift in funding from other accounts to E&HR would not necessarily result in a real increase to NSF's STEM education programs (particularly if the reductions were to R&RA, which could presumably offset such reductions by limiting its contributions to co-funded programs). Further, reductions to other accounts may limit NSF's ability to meet Congress's non-STEM education priorities.
- *Increase the R&RA contribution to E&T*—Another way to increase overall funding for education at the NSF would be to increase the R&RA contribution to E&T activities through increased contributions to co-funded activities. Congress could do this with or without overall increases to the NSF and R&RA; however, if R&D funding does not also increase at the rate of inflation, purchasing power may be lost. Increasing R&RA contributions to E&T activities may deepen the integration of these complementary NSF missions or may put pressure on research activities.

- **Decrease funding for STEM education at NSF.** If Congress seeks to capture savings from the NSF budget or prioritize STEM education activities at other federal agencies, it could reduce funding for NSF STEM education programs. To this end, Congress has at least two options. It could prioritize certain programs— for example, by either portfolio role or performance—and reduce others. Alternatively, it could reduce topline support for E&HR and limit the percentage of R&RA that may be used for E&T.
  - *Prioritization*—In general, prioritization of certain programs might result in savings, depending on the choices policy makers make about reductions and support for the programs it preserves. However, congressional prioritization of NSF's STEM education programs may challenge the foundation's historical autonomy, which many analysts see as essential to its scientific mission. Congress has at least two options for prioritizing NSF's STEM education programs: portfolio role and performance.

    - *By Portfolio Role*—Instead of treating NSF's education activities as a single function, Congress could separate NSF's STEM education R&D programs from its student and institutional aid programs. Congress could then establish differential funding rates for NSF's STEM education R&D and aid programs. For example, some analysts suggest that research funding is most efficient when provided in predictable incremental increases (as opposed to wide variations, which impose adjustment costs). However, demand for student and institutional aid tends to be affected by factors that vary—such as population size, general economic conditions, and state education budgets, among others.
    - *By Performance*—Congress could direct NSF to develop a framework for evaluating its STEM education programs. That framework could incorporate factors that reflect the importance of the program to NSF's mission, to the federal STEM education portfolio, to the constituencies served, to the field (e.g., intellectual merit), and to other congressional criteria. Policy makers at NSF and Congress could use this information to prioritize funding for NSF's STEM education programs. A performance approach to reductions could increase the effectiveness of NSF's STEM education programs. However, one of the challenges of the performance approach is that the criteria by which decisions are made may not reflect the full value of the programs and as a result, effective programs may be unintentionally terminated.
  - *Decrease topline funding for E&HR and limit the percentage of R&RA that may be used for E&T.* Given the important role played by federal funding for fundamental research—most of NSF-funded research is basic research— Congress could prioritize NSF's research activities over its education activities and continue assigning most of the foundation's funding growth to the research account. To the extent that NSF's STEM education programs are unique to the federal effort, this may affect the portfolio.
- **Use mechanisms other than the NSF to achieve federal STEM education goals.** Congress could meet federal STEM education goals in any number of ways, depending on what those goals are and the

policy strategies policy makers pursue. For example, Congress could increase the number of students who are interested in and prepared to study STEM subjects in college by increasing funding for Advanced Placement or other gifted student programs at ED. Other analysts may seek to increase general federal student aid (e.g., Pell program), which also serves STEM students (along with all others) but does not create incentives for students to pursue degrees in certain fields. Some policy makers may prefer to leave such matters to state and local governments to decide.

- **Decrease funding for STEM education across the portfolio.** If Congress seeks to prioritize other national concerns (e.g., national debt, defense, health care), it may choose to reduce funding for STEM education across the federal enterprise. Federal agencies may respond to reduced funding levels by limiting STEM education activities. Congress may wish to consider providing guidance to federal agencies to ensure that legislative priorities are maintained.

As Congress weighs these various options, it may be useful to consider the short-, medium-, and long-term impact of congressional funding choices on the entire federal STEM education portfolio, on the respective research and education missions of the NSF, and on the general policy purposes (e.g., advancement of the national STEM labor supply) these investments seek to serve.

APPENDIX A. CHARACTER CLASS DATA

Table A-1. NSF Funding by Character Class (Current) FY2003-FY2012 Actual, In Millions, Rounded

Class	FY 2003	FY 2004	FY 2005	FY 2006	FY 2007	FY 2008	FY 2009	FY 2010	FY 2011	FY 2012
R&D	$3,944	$4,166	$4,098	$4,240	$4,407	$4,506	$4,729	$5,488	$5,537	$5,705
NIA	$524	$545	$559	$624	$675	$722	$843	$584	$576	$593
E&T	$901	$941	$824	$783	$803	$856	$897	$900	$800	$807
Total	$5,369	$5,652	$5,481	$5,646	$5,884	$6,084	$6,469	$6,972	$6,912	$7,105

Source: CRS calculations based on data provided by the NSF.

Table A-2. NSF Funding by Character Class (2005 Constant) FY2003-FY2012 Actual, In Millions, Rounded

Class	FY 2003	FY 2004	FY 2005	FY 2006	FY 2007	FY 2008	FY 2009	FY 2010	FY 2011	FY 2012
R&D	$4,175	$4,302	$4,098	$4,100	$4,139	$4,137	$4,286	$4,924	$4,866	$4,923
NIA	$555	$562	$559	$603	$634	$662	$764	$524	$506	$512
E&T	$954	$972	$824	$757	$754	$786	$813	$808	$703	$696
Total	$5,684	$5,836	$5,481	$5,460	$5,527	$5,585	$5,863	$6,256	$6,075	$6,131

Source: CRS calculations based on data provided by the NSF.

Appendix B. Appropriations Data

Table B-1. NSF Appropriations by Account (Current) FY2003- FY2012 Actual, In Millions, Rounded

Account	FY 2003	FY 2004	FY 2005	FY 2006	FY 2007	FY 2008	FY 2009	FY 2010	FY 2011	FY 2012
R&RA	$4,144	$4,388	$4,328	$4,449	$4,758	$4,853	$5,152	$5,615	$5,608	$5,758
E&HR	$846	$850	$750	$700	$696	$766	$846	$873	$861	$831
MREFC	$179	$184	$165	$234	$166	$167	$161	$166	$125	$198
AOAM	$189	$219	$223	$247	$248	$282	$294	$300	$299	$299
NSB	$3	$2	$4	$4	$4	$4	$4	$4	$4	$4
OIG	$9	$9	$10	$11	$12	$12	$12	$14	$14	$14
Total	$5,369	$5,652	$5,481	$5,646	$5,884	$6,084	$6,469	$6,972	$6,913	$7,105

Source: CRS calculations based on data provided by the NSF.

Table B-2. NSF Appropriations by Account (2005 Constant) FY2003- FY2012 Actual, In Millions, Rounded

Account	FY 2003	FY 2004	FY 2005	FY 2006	FY 2007	FY 2008	FY 2009	FY 2010	FY 2011	FY 2012
R&RA	$4,387	$4,530	$4,328	$4,303	$4,470	$4,455	$4,670	$5,038	$4,929	$4,969
E&HR	$895	$878	$750	$677	$653	$703	$766	$783	$757	$717
MREFC	$190	$190	$165	$226	$156	$153	$146	$149	$110	$171
AOAM	$201	$226	$223	$239	$233	$259	$267	$269	$263	$258
NSB	$3	$2	$4	$4	$3	$4	$4	$4	$4	$4
OIG	$9	$10	$10	$11	$11	$11	$11	$13	$12	$12
Total	**$5,684**	**$5,836**	**$5,481**	**$5,460**	**$5,527**	**$5,585**	**$5,863**	**$6,256**	**$6,075**	**$6,131**

Source: CRS calculations based on data provided by the NSF.

End Notes

[1] Another source of STEM labor in the United States is immigration. For more information about foreign STEM workers, see CRS Report R42530, *Immigration of Foreign Nationals with Science, Technology, Engineering, and Mathematics (STEM) Degrees*, by Ruth Ellen Wasem.

[2] Liana Christin Landivar, "The Relationship Between Science and Engineering Education and Employment in STEM Occupations," *American Community Survey Reports* (ACS-23), U.S. Census Bureau, Economics and Statistics Administration, U.S. Department of Commerce, September 2013. See also CRS Report R43061, *The U.S. Science and Engineering Workforce: Recent, Current, and Projected Employment, Wages, and Unemployment*, by John F. Sargent Jr.

[3] For more information about federal STEM education activities, see CRS Report R42642, *Science, Technology, Engineering, and Mathematics (STEM) Education: A Primer*, by Heather B. Gonzalez and Jeffrey J. Kuenzi.

[4] Funding at the NSF is distributed between accounts that primarily support research and accounts that primarily support STEM education. However, these missions are highly interrelated. STEM education ultimately enables the conduct of research, which by its very nature is often educational. Nevertheless, funding for these two missions supports different activities.

[5] The post-sequestration, post-rescission FY2013 current plan funding level for NSF's main education account is $833.3 million. This is slightly higher than the FY2012 actual funding level of $830.5 million. The FY2013 current plan funding level may differ from the final, FY2013 actual funding level. Current plan funding levels are typically estimated.

[6] NSF adheres to Office of Management and Budget, *OMB Circular A-11*, for its character class definitions. For more information, see http://www.whitehouse.gov/omb/circulars_a11_current_year_a11_toc.

[7] As the purpose of this report is to examine NSF's investment activities, it excludes NIA trends.

[8] The America COMPETES Act (P.L. 110-69) increased authorized funding levels for targeted accounts at the NSF, National Institutes of Standards and Technology laboratories and construction, and the Department of Energy's Office of Science. For more information see CRS Report R41819, *Reauthorization of the America COMPETES Act: Selected Policy Provisions, Funding, and Implementation Issues*, by Heather B. Gonzalez; and CRS Report R41951, *An Analysis of Efforts to Double Federal Funding for Physical Sciences and Engineering Research*, by John F. Sargent Jr.

[9] For example, in FY2005, both R&RA and E&HR experienced reductions. The decrease in R&RA was $59.4 million, or 1.4% percent below the FY2004 level of $4.388 billion. The following year (FY2006) Congress returned the R&RA account to the FY2004 level and increased beyond it. By contrast, the FY2005 decrease to E&HR was $99.7 million, or 11.7% of the FY2004 level ($849.9 million). Congress did not return E&HR funding to FY2004 levels until FY2009.

[10] One influential example of this argument is laid out in National Academy of Sciences, National Academy of Engineering, and Institute of Medicine, Committee on Prospering in the Global Economy of the 21st Century: An Agenda for America Science and Technology, and Committee on Science, Engineering, and Public Policy, *Rising Above the Gathering Storm: Energizing and Employing America for a Brighter Economic Future*, National Academies Press, 2007, http://www.nap.edu/catalog/11463.html.

[11] Richard Freeman, "The Market for Scientists and Engineers," *NBER Reporter*, no. 3 (Summer 2007), pp. 6-8, http://www.nber.org/reporter/2007number3/freeman.html.

[12] Ron Hira, "U.S. Policy and the STEM Workforce System," *American Behavioral Scientist*, v. 53, no. 7 (March 2010), pp. 949-961.

[13] Testimony of Institute for the Study of International Migration Director of Policy Studies B. Lindsay Lowell, in U.S. Congress, House Committee on the Judiciary, Subcommittee on Immigration Policy and Enforcement, *"STEM" the Tide: Should America Try to Prevent an Exodus of Foreign Graduates of U.S. Universities with Advanced Science Degrees?*, hearings, 112th Cong., 1st sess., Serial No. 112-64, October 5, 2011, http://judiciary hear_10052011_2.html.

[14] B. Lindsay Lowell and Harold Salzman, *Into the Eye of the Storm: Assessing the Evidence on Science and Engineering Education, Quality, and Workforce Demand*, Urban Institute, October 2007, http://www.urban.org/ publications/411562.html.

[15] Anthony P. Carnevale, Nicole Smith, and Michelle Melton, *STEM: Science, Technology, Engineering, and Mathematics*, Georgetown University Center on Education and the Workforce, October 20, 2011, p. 7, http://cew.georgetown.edu/STEM/.

[16] Ibid., p. 75.

[17] Executive Office of the President, National Science and Technology Council, Committee on STEM Education, Fast-Track Action Committee on Federal Investments in STEM Education, *The Federal Science, Technology, Engineering, and Mathematics (STEM) Education Portfolio*, December 2011, p. 68, http://www.whitehouse.gov/sites/default/files/ microsites/ostp/costem__federal_stem_education_portfolio_report.pdf.

[18] U.S. Government Accountability Office, Science, Technology, Engineering, and Mathematics Education: Strategic Planning Needed to Better Manage Overlapping Programs Across Multiple Agencies, GAO-12-108, January 2012, pp. 27-29, http://gao.gov/products/GAO-12-108.

[19] Executive Office of the President, National Science and Technology Council, Committee on STEM Education, Fast-Track Action Committee on Federal Investments in STEM Education, *The Federal Science, Technology, Engineering, and Mathematics (STEM) Education Portfolio*, December 2011, http://www.whitehouse.gov/sites/default/files/ microsites/ostp/costem__federal_stem_education_portfolio_report.pdf.

[20] Based on NSF briefing of CRS and GAO staff, January 21, 2011.

[21] See National Science Foundation, *FY2013 Budget Request to Congress*, February 13, 2012, p. Summary Tables-5, http://www.nsf.gov/about/budget/fy2013/index.jsp.

[22] For example, see Senator Tom Coburn, *The National Science Foundation: Under the Microscope*, April 2011, p. 54, http://coburn.senate.gov/public//index.cfm?a=Files.Serve& File_id=2dccf06d-65fe-4087-b58d-b43ff68987fa.

[23] The GAO found a total of 82 potentially duplicative teacher quality programs at 10 federal agencies. The auditing agency indicates that nine of these programs were at the NSF. See, U.S. Government Accountability Office, *Opportunities to Reduce Duplication in Federal Teacher Quality Programs* (GAO-11-510T), April 13, 2011, http://www.gao.gov /products/GAO-11-510.

[24] Executive Office of the President, National Science and Technology Council, Committee on STEM Education, Fast-Track Action Committee on Federal Investments in STEM Education, *The Federal Science, Technology, Engineering, and Mathematics (STEM) Education Portfolio*, December 2011, p. 37, http://www.whitehouse.gov/sites/default/files/ microsites/ostp/costem__federal_stem_education_portfolio_report.pdf.

[25] U.S. Government Accountability Office, *Science, Technology, Engineering, and Mathematics Education: Strategic Planning Needed to Better Manage Overlapping Programs Across Multiple Agencies*, GAO-12-108, January 2012, pp. 20-21, http://gao.gov/products/GAO-12-108.

[26] This is, for example, part of the rationale for Administration-proposed program consolidations at the Department of Education. For more information about the Administration's planned changes at ED, see CRS Report R41355, *Administration's Proposal to Reauthorize the Elementary and Secondary Education Act: Comparison to Current Law*, by Rebecca R. Skinner et al.

[27] This would depend on how the grants to states were structured. Federal policy makers could still attempt to establish a national STEM education agenda by making receipt of consolidated program funds contingent on meeting certain defined national goals. However, some states may reject such efforts as overly prescriptive.

[28] For example, the National Aeronautics and Space Administration (NASA) has both unique workforce needs (e.g., astrobiologists) and unique assets that it can bring to the national STEM education effort (e.g., teaching from space).

[29] GAO states that "over 90% of STEM education programs that reported administrative costs estimated having administrative costs lower than 10% of their total program costs." U.S. Government Accountability Office, *Science, Technology, Engineering, and Mathematics Education: Strategic Planning Needed to Better Manage Overlapping Programs Across Multiple Agencies*, GAO-12-108, January 2012, p. 22, http://gao.gov/products/GAO-12-108.

[30] For example, programs that appear duplicative by some measures (e.g., target group), may have different intangible assets that could impact program implementation and outcomes.

[31] For example, the December 2011 NSTC inventory of federal STEM education programs suggested that there was room for improvement in their management and stated that "the primary issue [instead of duplication] is how to strategically focus the limited federal dollars available within the vast landscape of opportunity so they will have the most significant impacts possible in areas of national priority." (See, NSTC report, p. 37.) The GAO concluded similarly in its January 2012 report and recommended that the NSTC draft a federal STEM education strategy plan, and that NSTC should also develop policies to ensure agencies comply with the national plan. In particular, the GAO recommended that NSTC should develop (1) guidance for agencies on how to incorporate STEM education efforts into agency performance plans; (2) a framework for how agencies will be monitored to ensure they collect and report on strategic plan goals; and (3) guidance to help agencies determine the types of evaluations that may be feasible and appropriate for different types of STEM education programs. Additionally, GAO recommended that NSTC should work with agencies to identify programs that might be candidates for consolidation or elimination. (See, GAO-12-108, p. 31).

[32] Executive Office of the President, National Science and Technology Council, Committee on STEM Education, Federal Coordination in STEM Education Task Force, *Coordinating Federal Science, Technology, Engineering, and Mathematics (STEM) Education Investments: Progress Report*, February 2012, http://www.whitehouse.gov/sites/default/files/microsites/ostp/nstc_federal_stem_education_coordination_report.pdf.

[33] Ibid., p. 13.

[34] Ibid., p. 17.

[35] Executive Office of the President, National Science and Technology Council, Committee on STEM education, *Federal Science, Technology, Engineering, and Mathematics (STEM)*

Education: 5-Year Strategic Plan, May 2013, p. 8, http://www.whitehouse.gov/sites/default /files/microsites/ostp/stem_stratplan_2013.pdf.

[36] Ibid., p. 16.

[37] This proposal is discussed more fully in the section titled, "Obama Administration Proposal to Reorganize the Federal STEM Education Effort in FY2014."

[38] Testimony of Office of Science and Technology Policy Director John Holdren, in U.S. Congress, Senate Committee on the Budget, *Silo Busting: Effective Strategies for Government Reorganization*, hearings, 113th Cong., 1st sess., May 16, 2013.

[39] H.Rept. 113-171, p. 59.

[40] House Committee on Science, Space, and Technology, Ranking Member Eddie Bernice Johnson, "Committee Discusses Proposed Reorganization of STEM Education Programs," press release, June 4, 2013, http://democrats.

[41] Testimony of Office of Science and Technology Policy Director John Holdren, in U.S. Congress, Senate Committee on the Budget, *Silo Busting: Effective Strategies for Government Reorganization*, hearings, 113th Cong., 1st sess., May 16, 2013.

[42] Ibid.

[43] The House report on Energy and Water Development Appropriations for FY2014 accepts some of the proposed changes. (See H.Rept. 113-135, p. 86.) However, both the House and Senate Appropriations Committee reports on Commerce, Justice, Science, and Related Agencies Appropriations reject the proposed reorganization in FY2014, except in specific instances. (See H.Rept. 113-171, p. 8, and S.Rept. 113-78, pp. 102-103.)

[44] House Committee on Science, Space, and Technology, "STEM Education: The Administration's Proposed Reorganization," hearings, 113th Cong., 1st sess., June 4, 2013, http://science.

[45] S.Rept. 113-78, p.103.

[46] Jeffrey Mervis, "An Invisible Hand Behind Plan to Realign U.S. Science Education," *Science*, vol. 341 (July 26, 2013), pp. 338-341. See also, "Science Education Overhaul," *Chemical & Engineering News* (June 17, 2013), pp. 25- 28.

[47] For more information on the effort to establish a federal STEM education strategy, see the section of this report titled, "A Federal STEM Education Strategy."

[48] In addition to the Mathematics and Science Partnership program, NSF's *FY2014 Budget Request to Congress* includes funding for the K-16 Mathematics Initiative. The K-16 Mathematics Initiative is a joint NSF-ED project focused on student learning of mathematics at the K-16 levels.

[49] U.S. Congress, House Committee on Appropriations, *Commerce, Justice, Science, and Related Agencies Bill, 2012*, report together with minority views to accompany H.R. 2596, 112th Cong., 1st Sess., H.Rept. 112-169, July 20, 2011 (Washington, DC:GPO 2011), pp. 84-85.

[50] More information about this effort is available at http://successfulstemeducation.org/.

[51] For example, some STEM education advocates argue that early undergraduate research opportunities are critical learning experiences and important to the STEM educational process.

In: National Science Foundation
Editor: Leslie Murphy
ISBN: 978-1-63321-972-4
© 2014 Nova Science Publishers, Inc.

Chapter 3

U.S. NATIONAL SCIENCE FOUNDATION: EXPERIMENTAL PROGRAM TO STIMULATE COMPETITIVE RESEARCH (EPSCoR)*

Christine M. Matthews

SUMMARY

The Experimental Program to Stimulate Competitive Research (EPSCoR) of the National Science Foundation (NSF) was authorized by Congress in 1978, partly in response to concerns in Congress and the concerns of some in academia and the scientific community about the geographic distribution of federal research and development (R&D) funds. It was argued that there was a concentration of federal R&D funds in large and wealthy states and universities, and that the continuation of such funding patterns might ensure a dichotomy between the "haves" and "have-nots."

EPSCoR began in 1979 with five states and funding of approximately $1.0 million. Currently, EPSCoR operates in 29 jurisdictions, including 27 states and the Commonwealth of Puerto Rico and the U.S. Virgin Islands. To 2006, the NSF had invested approximately $920.0 million in EPSCoR programs and activities. When established, it operated solely in the NSF. EPSCoR was expanded in the

* This is an edited, reformatted and augmented version of a Congressional Research Service publication RL30930, prepared for Members and Committees of Congress dated February 28, 2012.

mid-1980s and early 1990s; by 1998, seven other agencies had established EPSCoR or EPSCoR-like programs.

EPSCoR is a university-oriented program, with the goal of identifying, developing, and utilizing the academic science and technology resources in a state that will lead to increased R&D competitiveness. The program is a partnership between NSF and a state to improve the R&D competitiveness through the state's academic science and technology (S&T) infrastructure. Eventually, it is hoped that those states receiving limited federal support would improve their ability to compete successfully for federal and private sector funds through the regular grant system.

On November 18, 2011, President Barack Obama signed into law the Commerce, Justice, Science, and Related Appropriations Act, FY2012, P.L. 112-55. The law provides, among other things, funding for the NSF. The law provides a total of $7,033.1 million for the NSF in FY2012, $173.2 million above the FY2011 enacted level. Included in the total funding for NSF is $150.9 million for EPSCoR, approximately $5.5 million above the FY2011 level. The FY2012 appropriation for EPSCoR supports a portfolio of three complementary investment strategies—research infrastructure improvement ($110.0 million), co-funding ($39.4 million), and outreach and workshops ($1.5 million). It is anticipated that approximately 24.0% of the funding for EPSCoR in FY2012 will be used for new research awards. The remaining will be directed at providing support for grants made in previous years.

The Administration's FY2013 budget request for the NSF is $7,373.1 million, a 4.8% increase ($340.00 million) over the FY2012 estimate of $7,033.1 million. Included in the request total is $158.2 million for EPSCoR, an increase of 4.8% over the FY2012 estimate. Approximately 40.0% of the requested funding for EPSCoR in FY2013 will be directed toward new awards. The balance will support continuing awards made in prior years.

Background

The Experimental Program to Stimulate Competitive Research (EPSCoR) of the National Science Foundation (NSF) was authorized by Congress in 1978,[1] partly in response to concerns from Congress and from some of those in academia and the scientific community about the geographic distribution of federal research and development (R&D) funds.[2] Additional concerns resulted from the practice of congressional directed spending[3]—allocating funds for specific institutions or research projects.[4] Historical data revealed that there was a concentration of federal R&D funds in large and wealthy states and universities, and that the continuation of such funding patterns might ensure a dichotomy between the "haves" and "have-nots."[5]

As designed, EPSCoR is to help achieve broader geographical distribution of R&D support by improving the research infrastructure of those states that historically have received limited federal R&D funds. While these states fall outside of the top 10 states in receipt of federal R&D support, according to the NSF, they have "demonstrated a commitment to improve the quality of science and engineering research and education conducted at their universities and colleges."[6] The premise of the program is that "academic research activity underpins every state's overall competitiveness."[7] James Savage, writing in *Funding Science in America*, describes EPSCoR's creation as a type of "affirmative action program designed to aid less successful states and their universities in their competition for federal research funds."[8] W. Henry Lambright, Director, Center for Environmental Policy and Administration, Syracuse University, stated that "EPSCoR was not intended as an entitlement, but rather as a catalyst."[9] Lambright noted further that EPSCoR had a "troubled birth," having been rejected in its first vote by the National Science Board, the policy-making arm of NSF.[10] In order to win approval, the program had to be modified, expressing values consistent with those of the NSF: "... merit, with the emphasis on an institution, the university."[11]

EPSCoR began in 1979 with five states and funding of approximately $1.0 million. Currently, EPSCoR operates in 29 jurisdictions, including 27 states and the Commonwealth of Puerto Rico and the U.S. Virgin Islands. (*See Figure 1* for the participating jurisdictions.) To 2006, the NSF had invested approximately $920.0 million in EPSCoR programs and activities.[12] Currently, EPSCoR jurisdictions receive approximately 12% of all NSF research funding.[13] When established, it operated solely in the NSF. Congressional action led to its expansion in the mid-1980s and early 1990s, and by 1998, seven other agencies had established EPSCoR or EPSCoRlike programs.[14] This report is limited to a discussion of EPSCoR programs at the NSF.

EPSCoR Goals and Mission

Arden L. Bement, Jr., Former Director, NSF, stated that "EPSCoR is based on the premise that no one region and no one group of institutions has a corner on the market of good ideas, smart people, or outstanding researchers."[15] EPSCoR is a joint program of NSF and selected states and territories.

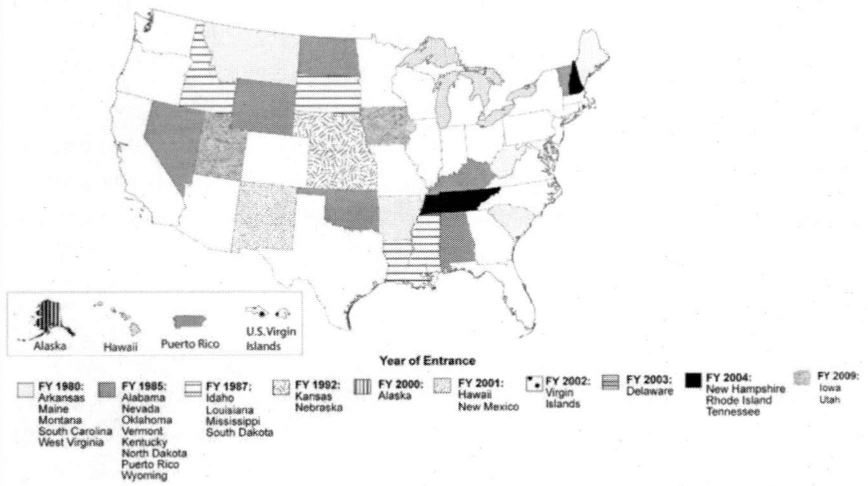

Source: NSF, 2009 EPSCoR Jurisdiction Map; http://www.nsf.gov/div/index.jsp?div=EPSC.

Figure 1. NSF EPSCoR Jurisdictions.

Its goal is to build competitive science by developing science and technology (S&T) resources through partnerships involving state universities, industry, government, and the federal R&D enterprise. The program is a partnership between the NSF and a state to improve the R&D competitiveness through the state's academic S&T infrastructure. The mission of EPSCoR is to raise the capability of a research institution or to assist in making a less-competitive institution more research intensive.[16] Eventually, EPSCoR supporters hope those states receiving limited federal support would gain some level of equity in competing for federal and private sector funds through the regular grant system.

The goal of the program as described by NSF is to

> Provide strategic programs and opportunities for EPSCoR participants that stimulate sustainable improvements in their R&D capacity and competitiveness, and to advance science and engineering capabilities in EPSCoR jurisdictions for discovery, innovation and overall knowledge-based jurisdictions. EPSCoR achieves its objectives by: (1) catalyzing key research themes and related activities within and among EPSCoR jurisdictions that empower knowledge generation, dissemination and application; (2)

> activating effective jurisdictional and regional collaborations among academic, government and private sector stakeholders that advance scientific research, promote innovation and provide societal benefits; (3) broadening participation in science and engineering by institutions, organizations and people within and among EPSCoR jurisdictions; and (4) using EPSCoR for development, implementation and evaluation of future programmatic experiments that motivate positive change and progression.[17]

In a prepared statement before the NSF EPSCoR 21st Annual Conference, Arden Bement stated that "Over the past 30 years, EPSCoR has evolved into a Program of Experimentation. It is a federal-state partnership that continues to show how to create and sustain robust infrastructures that support world-class research and education in science and engineering."[18]

OPERATION AND FUNDING

EPSCoR, while designed as a sheltered program, has been integrated into the performance of all NSF directorates.[19] Its grants are awarded on a competitive peer- or merit-reviewed basis. Proposals submitted vary, and come from academic, state, profit and nonprofit organizations, and individuals. Also, support is provided to cooperative programs among institutions in different EPSCoR states, or between a state's research institution and a primarily undergraduate institution. All principal investigators of NSF EPSCoR projects are required to be associated with research institutions, organizations, or agencies within the participating state. In addition, all of the projects must be designed to contribute to the research competitiveness of the colleges and universities in the particular state.

EPSCoR funding was not intended to replace existing federal, state, institutional, or private sector support, but to "... add specific value to the state's academic infrastructure not generally available through other funding sources."[20] Responsibility for operating the program rests within the individual states. A state is required to provide matching funds. An EPSCoR governing committee is established in each participating state to identify opportunities for EPSCoR awards. States devise strategies that allow them to adapt to vastly different federal funding environments. The programs are reviewed periodically by external panels and assessments are performed by independent organizations.

Data reveal that the 29 EPSCoR jurisdictions account for more than 20% of the U.S. population, about 25% of research institutions, and an estimated 16% of employed scientific and technical personnel. As a whole, these 29 jurisdictions receive approximately 13.6% of all NSF R&D funding.[21] In FY2010, NSF provided an estimated $145.4 million for EPSCoR activities, an increase of $12.4 million (9.3%) above the FY2009 level.[22] (See *Table 1* for funding levels of previous years.) Funding was provided through three complementary investment strategies— research infrastructure improvement grants, co-funding, and outreach and workshops.[23]

Table 1. Estimated EPSCoR Funding: FY2004-FY2013 (dollars in millions)

Fiscal Year	Research Infrastructure Improvement	Co-Funding	Outreach and Workshops	Funding Year Totals
FY2004	56.7	37.1	0.1	93.9
FY2005	58.6	34.3	0.1	93.0
FY2006	61.2	36.1	0.1	97.4
FY2007	65.3	35.9	0.1	101.3
FY2008	72.4	46.7	0.2	119.3
FY2009	119.9	40.5	.5	160.9
FY2010	118.9	44.8	1.5	165.2
FY2011	104.7	38.9	1.2	144.8
FY2012	110.0 (est.)	39.4 (est.)	1.5 (est.)	$150.9 (est.)
FY2013	116.2 (req.)	40.0 (req.)	2.0 (req.)	$158.2 (req.)
Total	$883.9	$393.7	$7.3	$1,284.9

Note: Funding levels provided by NSF Budget Office and NSF Office of Legislative and Public Affairs (e-mail communication, February 24, 2012) and U.S. National Science Foundation, FY2013 Budget Request to Congress, NSF12-036, Arlington, VA, February 13, 2012, p. IA-5.

NSF's current portfolio for EPSCoR includes three complementary investment strategies— research infrastructure improvement (RII) grants, co-funding of disciplinary and multidisciplinary research, and outreach and workshops. RII grants support S&T infrastructure improvements that have been designated by a governing committee in the EPSCoR state as essential to the state's future R&D competitiveness.[24] RII grants are of two types—RII Track 1 and RII Track 2. RII Track 1 grants are made to individual jurisdictions and are awarded up to $15.0 million for a period of up to 60 months. RII Track 2 grants are made to consortia of EPSCoR jurisdictions and are limited to a maximum of $2.0 million for up to 36 months. Examples of RII grants include startup funding for faculty research, faculty exchange projects with major research centers, development of nationally competitive high-performance computing capabilities, acquisition of state-of-the-art research instrumentation that is unavailable through the NSF's regular grant system, creation of graduate research training groups that encourage multidisciplinary experiences, developing linkages between industry and national laboratories, and development of programs to expand minority participation in science, engineering, mathematics, and technical disciplines. RII grants are the principal focus of the EPSCoR program. NSF funding for EPSCoR RII grants in the FY2012 appropriations is estimated at $110.0 million, a $5.3 million decrease from the FY2011 level. The Administration's request for RII in the FY2013 request is $116.2 million.

The co-funded grant mechanism encourages EPSCoR researchers and institutions to move into the mainstream of federal and private sector R&D support. Co-funding is an internal, cross-directorate, NSF funding mechanism.[25] Co-funding activities are applicable in the various directorates, the Office of Polar Programs, the Office of International Science and Engineering, the Office of Cyberinfrastructure, and the Office of Integrative Activities. Co-funding allows states to receive more support than would have available under EPSCoR alone.[26] Proposals supported are in areas that have been identified by the state's EPSCoR governing committee as critical to the future R&D competitiveness of the state or jurisdiction, and include, among other things, individual investigator-initiated research proposals and R&D encompassed by the various crosscutting and interdisciplinary programs in NSF.[27] To receive support for co-funding, a grant proposal must be, among other things, rated at or near the same level as the highly rated grants in the regular grant process. The FY2012 appropriation for co-funding is $39.4 million, a slight increase over the FY2011 level of $38.9 million. The Administration requests $40.0 million in FY2013 for co-funding.

The outreach and workshops funding mechanism of EPSCoR provides support for NSF program directors and relevant personnel to visit participating researchers in EPSCoR states and to further familiarize the states and researchers with NSF policies, practices, and programs. Also, it allows agency personnel to become more cognizant of the availability of resources within the states and their institutions. Outreach and workshops visits take two forms—those initiated by a host of an EPSCoR state or jurisdiction, and those initiated by NSF program officers. The visits may result in colloquia or seminars. It is NSF's contention that the contact provided by outreach visits will lead to an increase in both the quality and quantity of grant proposals submitted by participating states.[28] Funding for the outreach strategy in the FY2012 appropriation is estimated at $1.5 million, slightly above the $1.2 million enacted in FY2011. It is anticipated that funding in FY2012 will allow for expansion of activities that build regional and jurisdictional research infrastructure. The budget request for outreach and workshops in FY2013 is $2.0 million.

PROGRAM EFFECTIVENESS

In 1994, an evaluation of the EPSCoR program was conducted by the COSMOS Corporation.[29] The evaluation, released in May 1999, covered the period 1980-1994, and was designed to, among other things, determine whether participating states and their institutions had increased their share of federal R&D funds and to identify the EPSCoR program strategies that led to improvement of the states' and institutions' research competitiveness. The evaluation found that states' R&D competitiveness did improve and that EPSCoR had contributed to this competitiveness. The report stated that

> Based on the observed changes in federal and NSF shares, it can be concluded that the EPSCoR states' share of R&D funding did increase relative to the shares of the other states. To this extent, EPSCoR was associated with a lessening of the undue geographic concentration of R&D in the United States. Although the changes were small in absolute terms, this was a notable accomplishment in an era when research universities in nonEPSCoR states also were thriving and upgrading substantially.[30]

The report noted that NSF EPSCoR had facilitated the development of partnerships and linkages among institutions, state and federal government, and the private sector. It also revealed that while no state had graduated from EPSCoR (no longer receiving EPSCoR support), many EPSCoR research clusters had become fully competitive and no longer sought EPSCoR resources.[31] The evaluators determined that for colleges and universities in EPSCoR states, the cluster strategy may have been a more effective approach to improving research capability than that of supporting individual researchers or single research projects, a strategy used in the early years of EPSCoR.

An August 2006 report, *EPSCoR 2020:*[32] *Expanding State Participation in Research in the 21st Century—A New Vision for the Experimental Program to*

Stimulate Competitive Research (EPSCoR), found that while some participating states and jurisdictions had developed S&T capabilities that address national issues, they needed to progress at a faster pace in order to benefit more fully in a national research agenda.[33] The report stated that

> The task now is to accelerate the positive trends in building research infrastructure and capacity in the states, and to incorporate the expertise and capabilities of these states into the larger national research agenda. As we move into a time of doubling the federal commitment to basic research, it is particularly critical and appropriate to make a new commitment to the EPSCoR states that have been left behind in the S&T community... . It is imperative that all of NSF's science, engineering, and education programs adopt the concept of broadening geographical and cultural participation in NSF activities as part of their objectives. Programmatic planning should consider how best to include all states and their research institutions as potentially important S&T resources.[34]

The workshop that generated this report proposed that a more flexible RII grant program should be instituted. The position of the workshop participants was that since the states were heterogeneous, a "one-size program" should not be applied to the then 27 different jurisdictions.[35] Rather, the individual needs of the states should be a factor in determining the most effective strategy for infrastructure improvement. The current award structure is viewed as being no longer adequate for some jurisdictions to achieve a higher level of competitive science in some areas of research. It was proposed that RII grants be awarded for a period of up to five years, in the amount of $3.0 million-$5.0 million per year, per state or jurisdiction. The longer period of time for the grant would enable states to better implement their strategic plans. The increased level of funding would be related to the size of the jurisdiction and the extent of the "S&T transformative challenge."[36]

Another suggestion from the workshop was to place the EPSCoR program in NSF where its cross-directorate interactions would be maximized and integrated into all of the cutting edge initiatives of the agency. The FY2008 budget request for NSF did transfer the EPSCoR program from the Education and Human Resources Directorate to the Integrative Activities in the Research and Related Activities account. The FY2008 budget submission stated that "The relocation will allow the EPSCoR program greater leverage for improving the research infrastructure, planning complex agendas, and developing scientific and engineering talent for the 21st century."[37]

An additional recommendation of the workshop was for EPSCoR states and jurisdictions to become a "test bed" for new initiatives. The report noted that because EPSCoR has matured as a program, it should expand its research capacity by developing expertise in areas of national importance, such as homeland security and national defense, cyberinfrastructure, environmental observatories, coastal and ocean issues, and energy expenditures. With the proposed flexibility to the RII grant mechanism, the participating states and jurisdictions could pursue multiple strategies, such as support for transformative research and innovation that has been outlined in NSF's strategic plan.[38] The workshop participants maintained that "Developing expertise in topics of national importance will enhance success of proposals in other competitions."[39]

ISSUES

At the beginning of the EPSCoR program, some questioned the length of time required for a state to improve its research infrastructure. It was suggested to be five years, but that proved to be "... unrealistic, both substantively and politically."[40] Questions remain concerning the length of time states should receive EPSCoR support.[41] There are those in the scientific community who believe that some states and their institutions should assume more responsibility for building their research infrastructure and become less dependent on EPSCoR funds. They argue that some researchers and states have become comfortable with EPSCoR funding and are not being aggressive in graduating from the program. It continues to be called an experimental program after three decades, and no state has yet graduated from the EPSCoR program.[42] The issue of graduation from the program has generated considerable Congressional interest.

In August 2005, the NSF's Committee of Visitors (COV) released a review of the EPSCoR program for the period FY2000 through FY2004.[43] One of the issues in the August 2005 review was centered on determining when states would become independent of EPSCoR resources. Questions included What initiatives are there to promote graduation from EPSCoR and mainstreaming in the regular grant making process? What level of progress must a state achieve to justify that it is no longer eligible for EPSCoR resources? The COV admitted that graduation/progression from the EPSCoR program is a "challenging" issue and has been debated from the beginning of the program within NSF and among the various stakeholders and participating states and jurisdictions. The review determined that it has become necessary to revisit what it means to "graduate" from the program. Because of the importance in developing a mechanism or measure for graduation from the program, the COV proposed the creation of a dedicated EPSCoR Advisory Committee (external) that would make recommendations for both eligibility for and graduation from EPSCoR. The report stated that

> Clearly, a fixed definition of graduation would be a moving target, especially in an environment where jurisdictions are still being added to the EPSCoR family. The current Office Head has articulated a vision of "programmatic graduation/progression," which necessarily includes the evolution of the EPSCoR programs themselves as infrastructure continues to grow. This vision should be further developed, vetted, and eventually implemented.[44]

The issue of increasing the number of states seeking support through the program was addressed in the review. The COV noted that the increase in the number of eligible jurisdictions has strained limited resources. In FY2002, 5,595 proposals were received, and 1,511 awards were made with a funding rate of 27.0%. In FY2006, 7,037 proposals were received, and 1,489 awards were made with a funding rate of 21.0%.[45] The report stated that

> Given the likely budgetary constraints to be imposed on EPSCoR in the coming years, the program runs the danger of not being able to serve its core clientele with the limited funds available if the number of eligible states and institutions continues to increase. At some point, the Foundation must more fully address infrastructure, capacity, and geographic distribution in its other grant programs. One solution might be for the Foundation to reorganize some of its existing programs in order to create an EPSCoR-like program that used "institutional competitiveness" rather than "state competitiveness" as the primary definitional criterion for support.[46]

Additional issues and questions were included in the August 2005 review by the COV. The review found that the majority of EPSCoR programs were capacity building based (infrastructure). The COV proposed that the significant number of capacity building programs should be supplemented with "complementary programs for building capability and competitiveness."[47] The SBRC grant mechanism was cited as important to expanding the "competitiveness" building component of EPSCoR. Also, the COV report found that while EPSCoR's program portfolio was diverse, and included minority serving institutions, community colleges, and high schools, it was determined that EPSCoR jurisdictions should further strengthen the linkages between faculty at minority serving institutions and those at research intensive institutions.

An examination of the review process for large RII-type proposals concluded that the review process should be more rigorous. The COV proposed including site visits in the review process, and in enlarging the pool of reviewers in the scientific and technical areas proposed for research. The review noted that with the current, relatively small number of reviewers of EPSCoR programs, there is "insufficient injection of new viewpoints in the review process."[48] The report suggested that the pool of reviewers should be expanded by rotating in a minimum of 25.0% new reviewers each year. The report further proposed that EPSCoR management use the review model employed by NSF's Engineering Research Centers, Science and Technology Centers, and Science of Learning Centers.

Many of the questions posed by the EPSCoR COV following its review of the program are those that are being debated by the various stakeholders in the EPSCoR community. In particular, questions concerning the criteria used to determine when a state or jurisdiction graduates from the EPSCoR program may continue to receive Congressional attention during the 112th Congress.

CONGRESSIONAL ACTIVITY

On March 2, 2007, Senator Rockefeller introduced S. 753, EPSCoR Research and Competitiveness Act of 2007. S. 753 would have authorized appropriations for FY2008-FY2012 to NSF for EPSCoR in the following amounts: FY2008, $125.0 million; and FY2009-FY2012, $125.0 million and "... $125,000,000 multiplied by a percentage equal to the percentage by which the Foundation's budget request for such fiscal year exceeds the total amount appropriated to the Foundation for fiscal year 2008."[49] Language in the bill would have required the development of plans that allow EPSCoR states and jurisdictions to participate in NSF's Cyberinfrastructure Initiative and Major Research Instrumentation program. S. 753 would have required the NSF Director to obligate not less than 20.0% of the EPSCoR budget on co-funding projects that are ranked, by peer review, in the top 20.0% of all submitted grant proposals. Also, EPSCoR states and jurisdictions participating in the RII grant mechanism would have had to include in the proposals, partnerships with out-of-state research institutions. S. 753 was referred to the Senate Committee on Health, Education, Labor, and Pensions. It saw no further action.

On January 4, 2011, President Obama signed into law the America COMPETES Reauthorization Act, FY2010 (P.L. 111-358).[50] The law authorizes appropriations for the NSF from FY2011 through FY2013 (FY2011, $7,424.4 million; FY2012, $7,800.0 million; and FY2013, $8,300.0 million). The authorization does not provide any specific funding levels for EPSCoR. However, the legislation includes language stating that

- The NSF Director shall continue EPSCoR to help eligible states develop their research infrastructure that will make them more competitive for research funding. The program shall continue to increase as NSF funding increases.
- A cross-agency evaluation of EPSCoR and other federal EPSCoR-like programs will be conducted, examining accomplishments, management, investment, and metric-measuring strategies implemented by the different agencies directed at increasing the number of new investigators receiving peer-reviewed funding. The examination will also include the degree of broadening of participation, knowledge generation, application, and linkages to national research and development competitiveness.[51]

The legislation also directs the National Academy of Sciences to conduct a study on the EPSCoR program to determine, among other things, the effectiveness of each state program; recommendations for improvements for all participating agencies to reach EPSCoR goals; and an assessment of the effectiveness of participating states in using awards to develop and build their science and engineering infrastructure. This study of EPSCoR and EPSCoR-like programs will provide recommendations that may result in policy implications for various federal agencies.[52] It is anticipated that this study will be completed in August 2013.

On November 18, 2011, the President signed into law the Commerce, Justice, Science, and Related Agencies Appropriations Act, FY2012 (P.L. 112-55).[53] The law provided a total of $7,033.1 million for the NSF in FY2012, $120.5 million above the FY2011 enacted level of $6,912.6 million. EPSCoR is contained within the Integrative Activities account and was funded at $150.9 million in FY2012,[54] $4.1 million above the FY2011 level. The FY2012 appropriation supported a portfolio of three complementary investment strategies—research infrastructure improvement ($110.0 million), co-funding

($39.4 million), and outreach and workshops ($1.2 million)—for the 29 EPSCoR jurisdictions. The NSF indicated that approximately 24.0% of the funding for EPSCoR is to be used for new research awards in FY2012, with the balance providing continuing support for ongoing grants.

The Administration's FY2013 budget request for the NSF is $7,373.1 million, a 4.8% increase over the FY2012 estimate of $7,033.1 million. Included in the total is $158.2 million[55] for EPSCoR, an increase of 4.8% over the FY2012 level. NSF estimates that approximately 40.0% of the FY2013 requested funding will be made available for new awards. The balance will be directed at supporting awards made in prior years.

End Notes

[1] Initial support for EPSCoR was contained in P.L. 95-392 (H.Rept. 95-1265), Department of Housing and Urban Development-Independent Agencies Appropriation Act, 1979.

[2] House Committee on Science and Technology, Subcommittee on Science, National Science Board: Science Policy and Management for the National Science Foundation, 1968-1980, Report, 98th Cong., 1st Sess., January 1983, p. 121. See also Colwell, Rita, Director, National Science Foundation, Speech before the Tenth Anniversary of Coalition of EPSCoR States, Washington, DC, March 23, 1999, http://www.nsf.gov/news/speeches/colwell/rc90323epscor.htm, p. 2. The charter of the NSF directs the agency to, among other things, "... avoid undue concentration of such research and education." National Science Foundation Act of 1950, Section 3. 7(c). U.S. Congress, House Committee on Science and Astronautics, A Bill to Amend the National Science Foundation Act of 1950, 89th Cong., 2nd Sess., April 19, 20, and 21, 1966, pp. 2-3.

[3] Also known as Congressional earmarks.

[4] Congressional earmarking continues to be a concern in the scientific and academic community. See for example CRS Report RL34462, House and Senate Procedural Rules Concerning Earmark Disclosure, by Sandy Streeter, Dennis, Steven T., "Artfully Redefining Earmarks," Roll Call, January 12, 2009,pp.1, 22, and Brainard, Jeffrey and JJ Hermes, "Colleges' Earmarks Grow, Amid Criticism," The Chronicle of Higher Education, v. 54, March 28, 2008, p. A1.

[5] The historic concentration of federal R&D support to institutions has changed slightly. In FY2007, the first 100 institutions (in terms of receipt of federal R&D funds) accounted for 82.6% of the total federal R&D support for science and engineering to colleges and universities. In FY1996, the top 100 institutions commanded 83.4% of support; in FY1986, the proportion was 85.5%. National Science Foundation, Federal Science and Engineering Support to Universities, Colleges, and Selected Nonprofit Institutions, Fiscal Year 2007, NSF09-315, Arlington, VA, September 2009, Table 19, Federal Science and Engineering Support to Universities, Colleges, and Nonprofit Institutions, Fiscal Year 1996, NSF98-331, September 1998, Table B-4, and Federal Support to Universities, Colleges and Selected Nonprofit Institutions FY1986, NSF87-318, January 1988, Table B-9. An analysis of state profiles for FY2007 reveal that the top 10 states accounted for 57.4% of federally funded academic R&D, while the 10 states with the smallest share totaled 3%. The states with the

smallest share are all EPSCoR states. National Science Foundation, Science and Engineering State Profiles: 2006-2008, NSF10-302, Arlington, VA, November 2009, Summary Tables. State profiles are based on data for the 50 states, the District of Columbia, and Puerto Rico.

[6] National Science Foundation, Annual Report FY2002, Experimental Program to Stimulate Competitive Research, Education and Human Resources Directorate, Arlington, VA, October 23, 2002, p. 1.

[7] [National Computer Science Alliance] NCSA Access: EPSCoR Conference, http://access.ncsa.uiuc.edu/Stories/ 97Stories/EPSCoR.html, p. 1.

[8] Savage, James D., Funding Science in America: Congress, Universities, and the Politics of the Academic Pork, Cambridge University Press, New York, 1999, p. 61.

[9] Lambright, W. Henry, Syracuse University, Paper prepared for the American Association for the Advancement of Science, Workshop on Academic Research Competitiveness, Coeur d'Alene, Idaho, Building State Science: The EPSCoR Experience, October 1-3, 1999, http://www.aaas.org/spp/rcp/epscor/lambright.htm, p. 2.

[10] U.S. Congress, House Committee on Science and Technology, The National Science Board: Science Policy and Management for the National Science Foundation, 1968-1980, Report prepared by the Science Policy Research Division, Library of Congress for the Subcommittee on Science, Research and Technology, 98th Cong., 1st Sess., January 1983, p. 121.

[11] Lambright, W. Henry, Workshop on Academic Research Competitiveness, p. 3.

[12] National Science Foundation, EPSCoR 2020: Expanding State Participation in Research in the 21st Century—A New Vision for the Experimental Program to Stimulate Competitive Research (EPSCoR), A report to the National Science Foundation, August 2006, p. iii. The current 29 participating jurisdictions account for more 20% of the U.S. population, 18% of employed scientists and engineering, and approximately 25% of research institutions. Note: A summary of EPSCoR funding since its inception to the present will be available following the release of the FY2013 budget request. (E-mail communication with NSF's Office of Legislative and Public Affairs and the EPSCoR program office, January 19, 2012).

[13] Schopf, Jennifer, NSF, Office of Integrative Activities, EPSCoR-Regional Grants Conference, March 21, 2011, Nashville, TN, http://www.nsf.gov/bfa/dias/policy/docs/epscor_nash.pdf.

[14] Currently, EPSCoR and EPSCoR-like programs are operational in the National Aeronautics and Space Administration (NASA), Department of Energy (DOE), Environmental Protection Agency (EPA), Department of Defense (DOD), National Institutes of Health (NIH, Institutional Development Award, IDeA), and Department of Agriculture (USDA). In FY1993, congressional action led to the creation of the EPSCoR Interagency Coordinating Committee (EICC), with the purpose of improving coordination among the participating agencies and ensuring consistency in implementation of EPSCoR policies. NSF is the lead agency within the federal-wide effort, and an NSF official serves as chair and executive secretary of EICC. In FY1998, the Department of Commerce (DOC) established the Experimental Program to Stimulate Competitive Technology (EPSCoT). It was designed as a technology complement to EPSCoR-like programs. In FY2010, 6 agencies provided a total of approximately $449.9 million for EPSCoR/EPSCoR-like programs. (In FY2006, EPA discontinued releasing separate EPSCoR program solicitations). The estimated FY2010 EPSCoR budgets by agency are as follow: NSF, $145.4 million, DOE, $22.0 million, USDA, $26.2 million; NIH, $228.9 million; NASA, $25.0 million; and DOD, funding level not available.

[15] Written remarks of Arden L. Bement, Jr., Former Director, National Science Foundation, before the NSF EPSCoR Project Directors and Administrators Annual Meeting, Arlington, VA, August 11, 2008. (Bement resigned his position at NSF in May 2010 to become Director, Global Policy Research Institute, Purdue University. Subra Suresh was sworn in on October 18, 2010, as the 13th director of NSF. Suresh had been Dean, School of Engineering, Massachusetts Institute of Technology.)

[16] Approximately 30% of minority-serving institutions are in EPSCoR jurisdictions. This includes 50% of historically black colleges and universities, 69% of tribal colleges and universities, and 29% of Hispanic serving institutions.

[17] About EPSCoR, http://www.nsf.gov/od/oia/programs/epscor/about.jsp.

[18] Arden L. Bement, Jr., Former Director, NSF, "From the Top," Luncheon Remarks, NSF EPSCoR 21st Annual Conference, Arlington, VA, October 21, 2009.

[19] A report of a FY2006 workshop on EPSCoR directed that funding for EPSCoR be moved from the Education and Human Resources (EHR) Directorate to Integrative Activities (IA) of the Research and Related Activities Account (R&RA). It was determined that the activities, management, and cross-directorate interactions of EPSCoR would be more fully coordinated and maximized within the R&RA. Beginning in the FY2008, EPSCoR was transferred from the EHR to the IA account.

[20] National Science Foundation, EPSCoR Research Infrastructure Improvement Grant Program (RII), Program Solicitation, NSF 03-528, Arlington, VA, January 2003, p. 5.

[21] National Science Foundation, Federal Funds for Research and Development: Fiscal Years 2006-08, Detailed Statistical Tables, NSF10-303, Arlington, October 2009, VA, Table 84.

[22] Funding levels include support for EPSCoR in the EHR Directorate and also co-funding available through the Research and Related Activities Account.

[23] Different investment strategies existed prior to the transfer of EPSCoR from the EHR to IA.

[24] Eligibility to participate in the EPSCoR Research Infrastructure Improvement competition is restricted to those states that received 0.75% or less of the total NSF research funds to all sources within a state averaged over the most recent three-year period. If in a rare instance a single large NSF-funded national or international facility skews the data, an adjustment to the eligibility criteria could be made. National Science Foundation, EPSCoR Research Infrastructure Improvement Grant Program, Program Solicitation, NSF08-500, Arlington, VA, January 4, 2008, p. 5.

[25] Proposals can not be submitted directly for co-funding. This mechanism is primarily internal to NSF, with reviews conducted by the EPSCoR office and the managing program office of the grant proposal. EPSCoR co-funding is limited to awards made by principal investigators at the eligible NSF EPSCoR jurisdictions.

[26] For expanded discussion of co-funding, see, for example, U.S. Congress, House Committee on Appropriations, Department of Veterans Affairs and Housing and Urban Development, and Independent Agencies Appropriations for 2001, 106th Cong., 2nd Sess., April 4, 2001, pp. 60-63, and Annual Report FY2002, Experimental Program to Stimulate Competitive Research, pp. 3-4.

[27] These crosscutting programs include those supported by multiple directorates in NSF and also joint agency programs. They include the Faculty Early Career Development Program, Professional Opportunities for Women in Research and Education, Major Research Instrumentation, Information Technology Research, Small Business Technology Research, Materials Research Science and Engineering Centers, Research Experiences for Undergraduates, Integrative Graduate Education and Research Training, and Grant Opportunities for Academic Liaison with Industry competitions.

[28] EPSCoR Outreach http://www.nsf.gov/od/oia/programs/epscor/outreach.jsp.

[29] National Science Foundation, An REC-Sponsored Report on Evaluation, A Report on the Evaluation of the National Science Foundation's Experimental Program To Stimulate Competitive Research, NSF99-115, Arlington, VA, May 1999, 36 pp. NOTE: The evaluation covered the period 1980-1994.

[30] Ibid., pp. 19-20.

[31] "A cluster is a related group of research projects, often interdisciplinary, and awards are made to a cluster's principal investigator as well as to the component research projects. This strategy has led to the development of laboratories or centers, and not just the recruitment of cadres of new, research-oriented faculty in a number of EPSCOR universities. The cluster strategy also has been used to promote interdisciplinary collaboration among universities and between universities and industry." Ibid., p. 36.

[32] A workshop is planned for January 19-20, 2012 for EPSCoR 2030, with a report to follow at a later time. (E-mail communication with NSF's Office of Legislative and Public Affairs, January 11, 2012.)

[33] EPSCoR 2020: Expanding State Participation in Research in the 21st Century—A New Vision for the Experimental Program to Stimulate Competitive Research (EPSCoR), 15 pp.

[34] Ibid., p. iv.

[35] (At the time of this particular workshop, there were 27 jurisdictions. Currently, there are 29.) Speaking before the 18th Annual EPSCoR National Conference, Sherry Farwell, Head, NSF EPSCoR Office, stated that it is necessary to provide funding opportunities for EPSCoR jurisdictions that account for their "... inherent jurisdictional heterogeneity. That is, jurisdictions vary in their current capacities and capabilities, and hence in their relative positions on the trajectory leading to competitiveness." Written statement of Sherry Farwell, 18th EPSCoR National Conference, Puerto Rico, September 26, 2005.

[36] EPSCoR 2020: Expanding State Participation in Research in the 21st Century—A New Vision for the Experimental Program to Stimulate Competitive Research (EPSCoR), p. 9.

[37] National Science Foundation, FY2008 Budget Request to Congress, Arlington, VA, February 2007, p. IA-1.

[38] National Science Foundation, Investing in America's Future, Strategic Plan FY2006-2011, NSF06-48, Arlington, VA, September 2006, 19 pp. NOTE: NSF describes transformative research as "... research that has the capacity to revolutionize existing fields, create new subfields, cause paradigm shifts, support discovery, and lead to radically new technologies." National Science Board, Enhancing Support of Transformative Research at the National Science Foundation, NSB07-6, Arlington, VA, January 23, 2007, p. 3.

[39] EPSCoR 2020: Expanding State Participation in Research in the 21st Century, p. 13. For update of this workshop see for example National Science Foundation, Experimental Program to Stimulate Competitive Research, Update: The EPSCoR 2020 Scorecard, August 11, 2008, http://www.nsf.gov/od/oia/programs/epscor/presentations/ HNBUpdate08 1108. pdf.

[40] Lambright, W. Henry, Workshop on Academic Research Competitiveness, p. 4.

[41] What constitutes success and termination from the program was included in the talking points at the EPSCoR Annual Meeting, Coeur d'Alene, Idaho, October 24-27, 2011, "EPSCoR Through the Lens of NSF's Strategic Plan," Clifford Gabriel, Acting Director, Office of Integrative Activities, NSF.

[42] In April 2006, it was announced that Tennessee would begin the process of transitioning out of the EPSCoR program. (Reflections, v.1, July 2006, p.1.) However, this did not occur. Tennessee did not transfer out and continued in a status that allowed full participation in the

program. (The reason was that for a brief period of time, Tennessee was above the participation threshold level). E-mail communication with NSF's program office of EPSCoR, June 30, 2011.

[43] National Science Foundation, Education and Human Resources Directorate, Experimental Program to Stimulate Competitive Research (EPSCoR), Committee of Visitors, Final Report, August 16, 2005, 25 pp.

[44] Ibid., p. 21.

[45] National Science Board, Report to the National Science Board on the National Science Foundation's Merit Review Process, Fiscal Year 2006, NSB07-22, March 2007, p. 31.

[46] Ibid., p. 23.

[47] Experimental Program to Stimulate Competitive Research (EPSCoR), Final Report, p. 12.

[48] Ibid., p. 24.

[49] S. 753, EPSCoR Research and Competitiveness Act of 2007, Section 3.

[50] For expanded discussion of the America COMPETES Reauthorization Act see CRS Report R41819, Reauthorization of the America COMPETES Act: Selected Policy Provisions, Funding, and Implementation Issues, by Heather B. Gonzalez. Note: This act reauthorizes selected provisions from the 2007 America COMPETES Act (P.L. 110-69).

[51] P.L. 111-358, Section 517 (b).

[52] U.S. National Science Foundation, FY2013 Budget Request to Congress, p. IA-4.

[53] P.L. 112-55, H.R. 2112, Consolidated and further Continuing Appropriations Act, FY2012, H.Rept. 112-284.

[54] In the FY2012 appropriation, NSF estimates that the number of people impacted by EPSCoR will be 4,470—755 senior researchers, 75 postdoctorates, 475 graduate students, 615 undergraduate students, 515 K-12 teachers, 1,800 K12 students, and 235 other professionals. U.S. National Science Foundation, FY2013 Budget Request to Congress, p. IA-6.

[55] NSF estimates that the FY2013 budget request for EPSCoR will involve a total of 4,680 individuals—790 senior researchers, 80 postdoctorates, 495 graduate students, 645 undergraduate students, 540 K-12 teachers, 1,885 K-12 students, and 245 other professionals. Ibid.

In: National Science Foundation
Editor: Leslie Murphy
ISBN: 978-1-63321-972-4
© 2014 Nova Science Publishers, Inc.

Chapter 4

NATIONAL SCIENCE FOUNDATION: STEPS TAKEN TO IMPROVE CONTRACTING PRACTICES, BUT OPPORTUNITIES EXIST TO DO MORE*

United States Government Accountability Office

WHY GAO DID THIS STUDY

The NSF spends more than $400 million of its $7 billion annual budget acquiring goods and services in support of its mission to promote science and engineering. Much of this spending involves exploration activities in remote locations throughout the world, such as the Arctic and Antarctic. GAO examined the extent to which NSF uses key contracting practices in three phases of the acquisition process: (a) acquisition planning, (b) contract award, and (c) post-award contract monitoring. GAO selected and reviewed a nongeneralizable sample of 11 contracts or orders with at least $3 million in funding obligations for fiscal year 2011, which accounted for about 70 percent of NSF's total contract obligations for that year. Although all 11 contracts and orders received funding during fiscal year 2011, some were awarded more than 7 years ago. Some were awarded more recently. We reviewed each of the

* This is an edited, reformatted and augmented version of a United States Government Accountability Office publication, No. GAO-13-292, dated March 2013.

11 contracts to determine the extent to which they reflected the use of key contracting practices based on the Federal Acquisition Regulation, our prior work, and NSF-OIG findings. GAO also reviewed NSF contracting policies and met with NSF contracting and program officials.

WHAT GAO RECOMMENDS

GAO recommends that the Director of NSF (1) supplement existing guidance on acquisition planning to address the time needed for the early stages of the process, and (2) arrange for audits to be performed on major contracts, consistent with the terms of the memorandum of understanding with NSF-OIG. NSF agreed with the recommendations.

WHAT GAO FOUND

For the contracts GAO reviewed, the National Science Foundation (NSF) generally used key contracting practices in each of the three phases of the acquisition process, but the agency needs additional guidance on early acquisition planning as well as arrangements for contract audits. The three phases of the process and key practices are shown in the figure below:

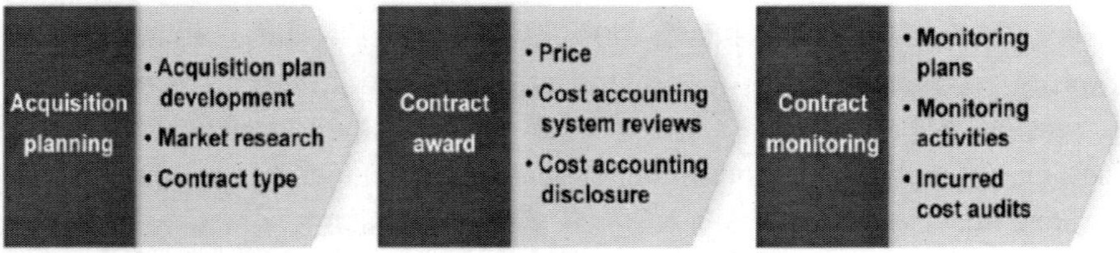

Source: GAO analysis of acquisition process.

Key Practices in the Acquisition Process.

The contracts GAO reviewed all involved some degree of acquisition planning, but NSF's guidance does not address appropriate time frames for early planning activities. Without such guidance, NSF contract and program officials said they could not convince their colleagues of the need to initiate early planning activities. Delays in these activities can lead to further delays later. For example, NSF had to extend one order on a non-competitive basis for more than a year to complete planning tasks for the follow-on order. In another case, the delayed award of an order compressed the data collection period for a report with firm deadlines, which could lead to higher overall costs. Further, having sufficient time for early planning may facilitate an increased use of lower risk contracting approaches.

Contract documentation showed that NSF generally followed key practices in the award phase. An NSF corrective action plan, in response to NSF's Office of Inspector General's (NSF-OIG) 2009 financial statement audits, clarifies the agency's procedures for reviewing contractors' accounting practices and financial disclosure statements to better align with key practices. Contract file documentation shows NSF improved in this area, with most of the negotiated contracts having documentation of accounting system reviews. Further, NSF generally documents price reasonableness determinations.

NSF updated its guidance and took steps to incorporate key contract monitoring practices. NSF-OIG's 2009 financial statement audits recommended that NSF obtain incurred cost submissions and audits for its largest cost-reimbursable contracts to ensure the validity of costs billed to NSF. Around the same time, the NSF-OIG and the NSF Office of the Director signed a memorandum of understanding (MOU) that provides a process for arranging for contract audits. Audits for one of the ocean drilling contracts completed in 2012 resulted in $1.5 million in recovered funds. The NSF Director and NSF-OIG have both identified additional audits of this contract as a top priority.

However, despite the terms of the MOU, and the agreement between NSF and the NSF-OIG on the need for further audits, arrangements have not been made to conduct additional audits of this contract for more recent fiscal years, according to officials. Similarly, despite requests from the contracting officer, NSF has not made arrangements for incurred cost audits for another large contract GAO reviewed.

ABBREVIATIONS

CAS	Cost Accounting Standards
DACS	Division of Acquisition and Cooperative Support
DCAA	Defense Contract Audit Agency
FAR	Federal Acquisition Regulation
IT	Information Technology
MOU	Memorandum of Understanding
NSF	National Science Foundation

NSF-OIG National Science Foundation Office of Inspector General
T&M Time and Materials

March 28, 2013

The Honorable Frank Wolf
Chairman, Subcommittee on Commerce, Justice, Science,
and Related Agencies
Committee on Appropriations
House of Representatives

Dear Mr. Chairman,

The National Science Foundation (NSF) spends more than $400 million annually acquiring the goods and services it needs to carry out its mission to promote science and engineering. Much of NSF's acquisition spending involves exploration activities in remote and austere locations throughout the world. Our work at several federal agencies has highlighted the importance of sound contracting practices—such as adequate time for planning, sufficient market research, and effective contract monitoring— as critical to a strong foundation for successful acquisition outcomes. But in recent years, we and others, including the NSF Office of the Inspector General (NSF-OIG), have identified various challenges regarding the contracting practices at NSF, particularly involving the agency's use of cost-type contracts.

You requested that we assess NSF's contracting practices. Specifically, our objective was to assess the extent to which NSF incorporates key contracting practices in the three major phases of the acquisition process: (a) acquisition planning, (b) contract award, and (c) post-award contract monitoring. To do so, we reviewed a nongeneralizable sample of 11 contracts and contract orders with funding obligations of at least $3 million each in fiscal year 2011, the latest year for which data were available when we began our work. We selected contracts and orders to reflect a mix of program offices and a variety of contract types, such as fixed-price and cost-reimbursement. The sample we used is not generalizable to the universe of all contracts at NSF, but the 11 contracts and orders represent 70 percent of the total dollars obligated by NSF on all its contracts and orders during 2011.[1] The 11 contracts and orders consisted of 4 contracts for which NSF used the negotiation process set forth in Part 15 of the Federal Acquisition Regulation (FAR) and 7 orders under existing contracts, which involved the use of streamlined procedures described in other parts of the FAR. Although all 11 contracts were active during the time of our review, some of the selected contracts were awarded more than 7 years ago—before NSF updated its contracting manual— and some more recently. As more fully described in appendix I, we determined the extent to which each of the contracts and orders in our sample reflected the use of selected key contracting practices. We selected the key practices based on requirements and principles in the FAR, practices identified in our prior work, or matters raised in NSF-OIG recommendations. In our view, each of the selected practices is critical to the foundation of a successful acquisition.

We conducted this performance audit from February 2012 to March 2013 in accordance with generally accepted government auditing standards. Those standards require that we plan and perform the audit to obtain sufficient, appropriate evidence to provide a reasonable basis for our findings and conclusions based on our audit objectives. We believe that the evidence obtained provides a reasonable basis for our findings and conclusions based on our audit objectives.

BACKGROUND

The acquisition process at federal agencies generally consists of three phases: (1) acquisition planning; (2) contract award; and (3) contract monitoring. Each phase involves a number of key activities, as shown in figure 1:

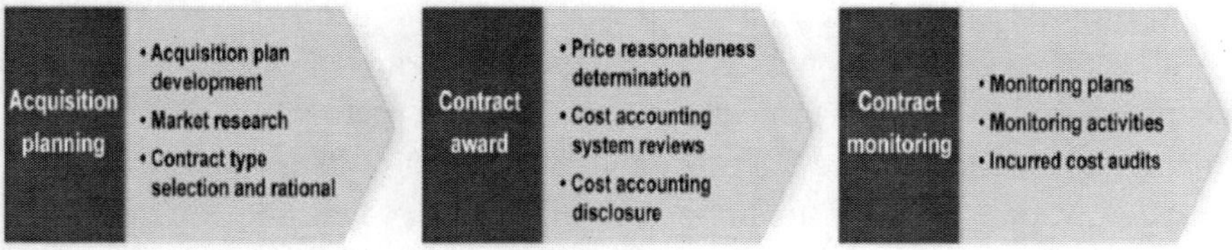

Source: GAO analysis of acquisition process.

Figure 1. Key Activities in the Acquisition Process.

In the acquisition planning phase, agencies establish their requirements and develop a plan to meet those requirements. Both program and contracting officials participate in acquisition planning activities.[2] During this phase, agencies conduct market research to determine what products or services are available and on what terms. They select a contracting approach best suited to the nature of the acquisition, addressing among other things, the availability of existing contracts, extent of competition required, and the most appropriate contract type, such as cost-reimbursable or fixed-price.

In the award phase, agencies solicit bids, quotes, or proposals from prospective vendors, depending on the contracting method selected. In negotiated acquisitions, they evaluate the submissions from vendors under established evaluation criteria in the solicitation and award a contract to the vendor representing the best value to the government, based on a combination of technical and cost factors. Agencies follow a similar process when ordering from the Federal Supply Schedule, where quotes from contractors are evaluated using stated evaluation criteria and orders are awarded to the contractor that would provide the best value and offers the lowest overall cost alternative.

In the contract monitoring phase, agencies engage in a range of activities intended to ensure that the contractor delivers according to the terms of the contract. These activities often are described in detail in a contract surveillance plan, sometimes called a quality assurance surveillance plan. For cost-reimbursement contracts, agencies may arrange for an audit of costs incurred by the contractor. These audits may be performed by entities such as the agency inspector general or the Defense Contract Audit Agency (DCAA).

NSF spends most of its annual budget of about $7 billion to fund grants to universities and other research entities, but the agency also spent more than $446 million in fiscal year 2011 acquiring goods and services in support of its mission. The largest of these acquisitions involved contracts for logistics support of scientific missions in the Arctic and Antarctica, as well as ocean-drilling projects in various locations. For these types of large-scale projects, NSF uses the negotiated contracting procedures of Part 15 of the FAR. NSF uses negotiated contracting methods for about 66 percent of its contract spending, as shown in figure 2. For another 32 percent of its contract spending, NSF uses a variety of more streamlined contracting methods allowed under the FAR. These include placing orders under Federal Supply Schedule contracts awarded by the General Services Administration or other pre-existing contracts.[3] Placing orders under existing contracts is often a more simplified approach than awarding a new contract. The remaining 1 percent or so of NSF contract spending is through various other methods, such as interagency agreements with the U.S. Navy for deep sea research vessel certification.

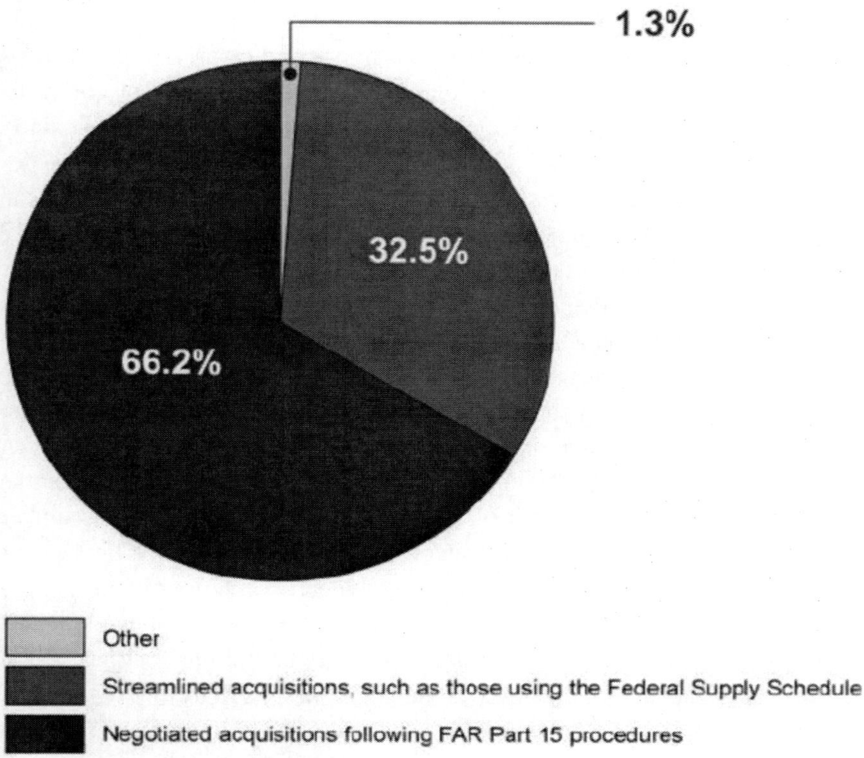

Source: GAO analysis of FPDS data.

Figure 2. Contracting Obligations by Acquisition Method, Fiscal Year 2011.

The Division of Acquisition and Cooperative Support (DACS) at NSF is responsible for the solicitation, negotiation, award, and administration of the agency's contracts for NSF's research facilities and major programs. DACS oversees NSF procurement systems, contracts policy, processes and guidance. This Division is under the Office of Budget, Finance, and Award Management which reports to the Office of the Director. The Office of Inspector General provides independent oversight of the agency's programs and operations, including contracts. The NSF-OIG is responsible for promoting efficiency and effectiveness in agency programs and for preventing and detecting fraud, waste, and abuse. By statute, the NSF-OIG is under the general supervision of the National Science Board and reports to the Board and Congress.

Much of NSF's contracting activity is for recurring needs, such as logistics support for its facilities in the polar regions, data collection, or surveys. For example, the National Survey of Recent College Graduates began in 1973 and continues today. In our prior work on acquisition planning practices, we found that documenting decisions, particularly when there is frequent staff turnover, is key to providing insight for subsequent contracts. Specifically, we found that documenting cost estimates is particularly important to help ensure the information is available when planning for follow-on contracts.[4] Incorporating lessons learned from prior acquisitions can help further refine requirements and strategies when planning for future acquisitions.

NSF officials must decide on a contract pricing arrangement for every contract or order. The major categories of pricing arrangements NSF uses are fixed-price, time-and-materials, and cost-reimbursement. Under a fixed-price contract, the government generally pays a firm price and may also pay an award or incentive fee related to performance. In a time and-materials contract or order, the government pays a set amount for every hour of service the contractor provides, plus the cost of any materials used. Because the number of hours to be provided is dependent on a number of factors, this type of contract requires an enhanced level of government oversight. When using a cost-reimbursement contract, the government agrees to reimburse all the allowable costs incurred by the contractor as prescribed in the contract. These types of contracts can be risky because the government agrees to pay for costs incurred regardless of the outcome achieved. Cost-type contracts that exceed certain dollar thresholds generally are subject to the cost allocation rules of the government's Cost Accounting Standards (CAS), and in these cases the contractor generally is required to disclose its cost accounting practices in a CAS Disclosure Statement. We previously reported on the use of cost-reimbursement contracts at several agencies, including NSF, finding that agencies frequently did not document why they selected this type of contract.[5]

Financial statement audits performed by an independent accounting firm on behalf of the NSF-OIG for fiscal years 2009 and 2010 identified significant deficiencies related to the use and monitoring of cost-reimbursement contracts at NSF. Specifically, the audits found that NSF did not ensure the adequacy of contractor accounting systems prior to award or the validity of costs incurred on the contract. In 2011, however, the same firm concluded that the concern had been addressed through the adoption of new policies and procedures. While we were conducting our audit work, NSF was in the process of conducting a self-assessment of its acquisition function in accordance with the Office of Management and Budget (OMB) Circular A-123. The agency also retained a consulting firm to review its self-assessment. In July 2012, the firm issued a report summarizing its findings. We did not assess the methodology, findings, or conclusions of either the NSF self-assessment or the consulting firm's review. In October 2012, NSF updated its contracting manual to incorporate a number of changes. For example, NSF reorganized the manual to align it with the FAR and added additional guidance to address the deficiencies identified in the financial audits. All of the contract activities in our review were subject to prior versions of the contracting manual.

ACQUISITION PLANNING PRACTICES CONDUCTED TO VARYING DEGREES, BUT NSF LACKS GUIDANCE ON TIME NEEDED FOR EARLY PLANNING ACTIVITIES

The NSF contract files we reviewed reflected the use of selected key acquisition planning practices to varying degrees, but the agency has not provided guidance on the time needed to complete early planning phase activities. Allowing sufficient time to plan procurements may facilitate an increased use of lower risk contracting vehicles by providing time for the contracting officer to consider including more fixed-priced elements. Our observations of the use of some of the key practices for acquisition planning activities are summarized in table 1, and explained in more detail below.

Table 1. Use of Key Practices in the Acquisition Planning Phase for the 11 Contracts/Orders Reviewed

	Key practices	Negotiated acquisitions	Streamlined acquisitions
Acquisition planning	Acquisition Plan. Comprehensive plan for fulfilling the agency need	◐	◐
	Market Research. Collecting and analyzing information about capabilities within the market	●	●
	Contract Type Rationale. Documents the agency's choice of contract type	◐	◐

Source: GAO analysis of NSF contract files.
● = Always or almost always used
◐ = Used to some degree
o = Used only to a limited degree or not at all

The acquisitions we reviewed all involved some degree of acquisition planning, but the time spent planning and the content of planning documents varied. Planning for the negotiated acquisitions ranged from a few months to more than 6 years, while many of the streamlined acquisitions in our sample had more abbreviated planning periods. Contracting and program officials responsible for one program office told us they often copy planning documents from predecessor orders to compensate for abbreviated planning periods. This practice, however, does not allow for incorporation of new guidance or changing contract requirements. In addition, some of the individual contract acquisition plans for the earlier contracts in our sample did not include details on how the agency planned to evaluate the proposals from competing vendors. Documenting a decision regarding the plan for proposal evaluation is an important component of the acquisition planning phase.

Contracting guidance at NSF does not identify the range of time needed to conduct acquisition planning activities for the types of acquisitions methods it employs. Currently, the guidance states that the process of acquisition planning should begin as soon as a program need is identified and it is determined that the need must be met through the use of resources from outside the government. The guidance does not provide any detail, however, on the expected range of time needed to conduct planning activities in the earliest stages of an acquisition when key documents are prepared, such as the statement of work and a cost estimate.[6] Acquisition planning usually occurs in three phases, and while NSF has established expected time frames for the latter stages of acquisition planning, the agency has not established such expectations for the earliest planning phase. Figure 3 depicts what we found at NSF.

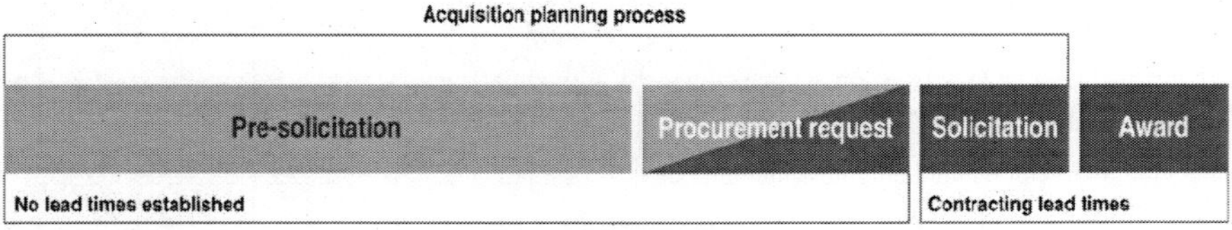

Source: GAO analysis of agency acquisition planning processes.

Figure 3. Acquisition Planning Phases.

Allowing sufficient time to plan procurements may provide agencies a better opportunity to clearly define contract requirements, outline source selection procedures, conduct market research to support competition, estimate costs, and consider opportunities for increased use of lower-risk contracting vehicles containing more fixed-priced elements. Conversely, the lack of sufficient time for planning may have adverse effects, such as unplanned delays. For example, NSF had to extend one streamlined order in our review on a non-competitive basis for more than a year and a half in order to complete planning tasks for the follow-on order. The contracting officer used the additional planning time to conduct the analysis needed to incorporate more fixed-priced elements into the new order. Planning for the earlier order did not include documentation of a price history analysis, which, according to contracting officials, may have helped expedite the follow-on planning and was likely due to short planning time frames for the earlier order. In another case, the delayed award of one of the orders in our review caused a compressed period for data collection for a report with firm deadlines. The schedule risk from these delays could lead to higher overall costs. Further, officials from two program offices told us that they would benefit from knowing an expected time range to complete early planning activities. For example, in the absence of guidance on the time needed to complete early planning activities, program and contracting officials responsible for NSF's largest contract told us they had difficulty convincing their colleagues of the appropriate time to initiate contract planning. They added that this acquisition required a number of changes before a follow-on contract could be awarded—some based on updates to the FAR and some based on internal decisions, including the use of a different source selection strategy.

Market research is a key element in the acquisition planning phase that provides insight into available sources for the acquisition and may provide information on estimated costs. We found evidence of market research in each of the acquisition plans we reviewed, though the link between the research conducted and the impact on the acquisition strategy was not always clear. For example, the acquisition plan for one streamlined acquisition noted concerns about the lack of offerors for past solicitations. The acquisition plan stated that NSF would use the Federal Supply Schedule and release the request for quotations to six potential offerors, but it did not address how market research impacted this decision. By contrast, NSF engaged in extensive planning for its Integrated Ocean-Drilling Program, including requirements development and market research to identify potential sources to support its mission. According to officials, this planning, which occurred over about a 5-year period, consisted of soliciting interest from more than 30 international institutions using various techniques such as market surveys and sources sought notices. NSF used this multi-year planning period to set up the funding and organizational infrastructure requirements of this complex international program.

All of the files we reviewed showed that during the planning phase agency officials had addressed how the contract would be priced. However, the planning documentation for the cost-reimbursable acquisitions in our review did not consistently include assessments of the additional risk and burden these high-risk contracts place on the agency or an assessment of the potential for firmer pricing in future acquisitions.[7] Knowing the risk of using a cost-reimbursable contract and identifying opportunities to use a less risky contract type after experience provides a basis for firmer pricing is a sound practice identified by our prior work, by the Department of Defense, and, more recently, in federal regulation.[8] Despite the risk associated with cost-type contracts, NSF contracting officials did not document their acknowledgment of this risk for an early contract for the ocean drilling program or whether they would attempt to minimize the future use of a cost-type contracts. Further, in a prior report we noted that NSF's procurements of data collection and analysis services for mandated surveys did not consider pricing history and whether there was a basis to transition to firmer pricing.[9] According to NSF officials, when re-awarding these types of survey procurements, staff will make an effort to identify tasks to convert to firmer pricing. In fact, a contracting officer responsible for the survey-related orders in our sample told us he has been conducting analysis to determine what tasks could be transitioned to a fixed-price contract type rather than a time-and-materials contract type. He stressed that some tasks are less suitable for fixed-pricing due to the unknowns and "what ifs" inherent in the work, but his goal is to incorporate fixed-pricing into 70 to 80 percent of each survey order. We identified examples of this transition to firm fixed-price elements in some of NSF's more recently awarded streamlined acquisitions.

NSF GENERALLY FOLLOWED KEY PRACTICES IN THE AWARD PHASE

Contract documentation for negotiated and streamlined acquisitions showed that NSF generally followed key practices in the award phase. Table 2 summarizes our findings based on the contracts and orders we reviewed.

Table 2. Use of Key Practices in Contract Award Phase for the 11 Contracts/Orders Reviewed

	Key practices	Negotiated acquisitions	Streamlined acquisitions
Award	Price reasonableness. Determination that the price or cost is fair and reasonable	●	●
	Cost Accounting SystemReview and pre-awardaudits. Help ensure contractor's cost accounting systems are adequate	●	n/a[a]
	CAS Disclosure Statement. Describes contractor's cost accounting practices and procedures	●	n/a

Source: GAO analysis of NSF contract files.

● = Always or almost always used

◐ = Used to some degree

○ = Used only to a limited degree or not at all

n/a = Practice is not required / applicable for this type of acquisition

[a]Cost accounting system reviews, pre-award audits, and CAS disclosure statement reviews are not applicable to streamlined acquisitions we reviewed.

Most of the contracts in our sample included price reasonableness determinations, as outlined in both federal regulation and NSF guidance current at the time of our review. For most of the streamlined acquisitions we reviewed, NSF documented reasonable price determinations, including an analysis of the contractor's proposed labor hours and the level of effort. In one case, contracting staff worked with an offeror to obtain lower labor rates that were more in line with the government cost estimate. These actions decreased the cost of the order by approximately 8 percent ($1.2 million).

In recent years, NSF has taken steps to address deficiencies related to accounting system and disclosure statement reviews identified in its fiscal year 2009 financial statement audits. Specifically, NSF clarified its CAS disclosure statement and accounting system review procedures to better align with sound practices identified by the NSF-OIG and in federal regulation.[10] Contract file documentation indicates that NSF has improved in this area, with most of the negotiated contracts we reviewed having documentation of more recent accounting system and CAS disclosure statement reviews, and the most recent contract having documentation of pre-award audits of all contractors in the competitive range. One of the earlier contracts did not have pre-award audits on file or an accounting system review prior to award. NSF officials told us that they did not think this requirement applied. In another earlier case, the contracting officer waived the requirement for a CAS disclosure statement adequacy determination prior to award with the expectation that the determination would be made shortly after award. However, NSF did not have documentation of the final disclosure statement adequacy determination.

NSF FOLLOWED PLANS FOR MONITORING PERFORMANCE, BUT INCURRED COSTS AUDITS WERE NOT COMPLETED REGULARLY

NSF updated its guidance and took steps to incorporate sound practices related to contract monitoring, but the agency has not made arrangements for audits of some of the larger contracts we sampled. Our findings are summarized in table 3.

Table 3. Use of Key Practices in Contract Monitoring Phase for the 11 Contracts/Orders Reviewed

Key practices		Negotiated acquisitions	Streamlined acquisitions
Contract monitoring	Monitoring Plans. Describe how the contract will be monitored	●	●
	Monitoring Activities. Help ensure the supplies and services acquired under the contract conform to requirements	●	●
	Incurred Cost Audits. Help ensure direct and indirect costs are allowable	◐	n/a[a]

Source: GAO analysis of NSF contract files.
● = Always or almost always used
◐ = Used to some degree
○ = Used only to a limited degree or not at all
n/a = Not required/applicable for this type of acquisition
[a]Incurred cost audits are not required for the type of streamlined acquisitions we reviewed.

Most of the contracts we reviewed included documentation of surveillance plans outlining how NSF would monitor contractor performance and costs, although one of the streamlined acquisitions did not have the surveillance documents called for in the acquisition plan. Further, we found evidence that at least some monitoring activities occurred for all the procurements we reviewed, though not always as specified in the monitoring plans or using deliverables described in the contract or order. For example, the acquisition plan for a large, information technology (IT) order states that the contractor shall provide "daily, weekly, and monthly progress reports" as well as an IT Management Plan and other ad hoc reports as required, with similar requirements reflected in the order. The contracting officer for this order was not aware of any daily progress reports for this order, and added that the monitoring process for these types of acquisitions depends on the quality of the contractor, noting that for some contracts with few performance issues, the monitoring is less rigorous. In another case, the contracting officer noted that despite the statement of work calling for a Quality Assurance Plan, such a plan would be too restrictive for an IT support contract due to the frequent changes in IT systems.

Our prior reports state that without consistent cost surveillance, such as through incurred cost audits, an agency may be exposed to the unnecessary risk of overpaying the contractor.[11] Further, NSF-OIG's fiscal year 2009 financial statement audits recommended that NSF obtain incurred cost submissions and audits for its largest cost-reimbursable contracts, depending on materiality and risk, to assure the validity of costs billed to NSF. In response, NSF updated its guidance on incurred cost audits and took the necessary steps to obtain incurred cost audits for its largest contract.

Around the same time, in August 2009, NSF-OIG and the NSF Office of the Director signed a memorandum of understanding (MOU) that provides procedures to ensure appropriate coordination between the NSF-OIG and NSF for the performance and funding of contract audits. The MOU indicates that the NSF-OIG will provide, within its resources, appropriated funds necessary to perform contract audits selected for its annual audit plan. The NSF-OIG solicits recommendations from NSF per the MOU and prioritizes its annual audit plan based on this input, its own needs, and a variety of risk factors. The MOU identifies the following factors the NSFOIG uses to prioritize contract audits: type of contract, materiality, whether NSF is the cognizant agency responsible for contractor oversight, known prior audit concerns, contract administration at other federal agencies, and whether NSF expects to continue to have a relationship with the contractor. For audits that NSF determines necessary that are not in the NSF-OIG audit plan, the MOU states that "NSF will obtain and fund the services of an outside auditor." The contracts branch officials told us that their first option is to ask the NSFOIG to obtain an audit, and if the NSF-OIG does not complete the contract audit, the branch tries to obtain alternative funding. NSF officials told us, however, that alternate funding requires approval at senior management levels, and contracting staff continue relying on the NSFOIG as the primary means for obtaining contract audits.

The NSF Director and NSF-OIG identified the need for incurred costs audits of an ocean drilling contract in our sample.[12] Despite the MOU, the agency has not made arrangements for these audits of the contract. Officials stated for earlier years of this contract, the Contracts Branch identified and provided funds for the contracting officer to initiate audits for this contract through DCAA. According to officials, an audit of a prime subcontractor for this contract resulted in $1.5 million in recovered funds. But at the time of our review, despite agreement on the importance of additional audits, the findings from the prior year's audits, and NSF's continued relationship with the contractor, the agency had yet to make arrangements to plan and fund incurred cost audits for more recent fiscal years for this contract, according to officials. Similarly, despite the contracting officer requesting incurred cost audits for another major contract in our review, the audit did not meet the NSF-OIG priorities. According to officials, NSF has not conducted or planned for audits on this contract. In addition, audits for another major contract we reviewed are not scheduled to be completed until fiscal year 2015, which is about two years after the contract expires.

In a recent report, we pointed out that timely closing of contracts, including completing any necessary incurred cost audits, can help the government limit its financial risk and possibly recover improper payments.[13]

CONCLUSION

Sound acquisition planning, including cost estimation and identification of the most cost-effective contract type, is important to establishing a strong foundation for successful outcomes for the millions of dollars NSF spends annually on acquisitions. Without sufficient planning time frames to develop acquisition plans that align with sound acquisition practices NSF may have a limited ability to develop a strong foundation for its acquisitions. How long the early acquisition planning activities should take is not covered in existing NSF guidance and will vary based on the complexity of the acquisition. However, without a clear understanding of the time frames needed for the early acquisition planning process, program officials may not know when to start planning or how long the planning will take, potentially increasing the likelihood of poorly prepared documents and contract delays. Better insights into when acquisition planning should begin would help ensure sufficient time to carry out the important acquisition planning activities that are designed to facilitate more successful outcomes.

When an acquisition involves substantial uncertainties and the agency deems a cost-type contract as the most appropriate vehicle, contract and program staff need to provide additional oversight to protect the government's interests. NSF has taken steps to address NSF-OIG recommendations to increase contract oversight. NSF has a management responsibility to ensure that adequate resources are available to enable contracting officers to determine that costs billed by contractors are allowable, through incurred cost audits or similar assessments. The process in place to ensure the necessary audits occur requires coordination between the NSF-OIG and the NSF Office of the Director; however, the process has not worked for some of the contracts we reviewed.

Further, the Contracts Branch continues to place a strong reliance on the NSF-OIG to provide the resources to obtain the audits. Without a process to ensure audits are conducted in cases when NSFOIG resources are not available, NSF exposes itself to unnecessary risk and cannot assure the validity of costs billed.

RECOMMENDATIONS FOR EXECUTIVE ACTION

We recommend that the Director of NSF take the following two actions:

- To help ensure good acquisition outcomes through comprehensive acquisition planning, direct DACS to supplement existing guidance on the time frames for acquisition planning to include a focus on the early stages.
- Consistent with the terms of the existing MOU with the Office of the Inspector General, take steps to arrange, and fund as necessary, timely audits of major contracts.

Agency Comments and Our Evaluation

We provided a draft of this report to NSF for review and comment. In written comments, NSF agreed with our recommendations. NSF also provided technical comments, which we incorporated as appropriate.

Sincerely yours,
William T. Woods
Director, Acquisition and Sourcing Management

Appendix I: Scope and Methodology

As requested by the Subcommittee on Commerce, Justice, Science, and Related Agencies, House Committee on Appropriations, we reviewed the National Science Foundation's (NSF) contracting practices. Specifically, we assessed the extent to which the NSF incorporates key contracting practices in the three major phases of the contracting process: (a) acquisition planning, (b) contract award, and (c) post-award contract monitoring.

Within each contracting phase, we focused our work on selected elements:

- Acquisition planning. We focused on the completeness and review of written acquisition plans, market research, contract type determinations, and time frames for planning. We selected these elements because they are critical to the successful planning of a contract and, in one case, had been identified in the past by the NSF Office of the Inspector General (NSF-OIG) as a potential concern.
- Contract award. We focused on cost and price analyses, cost accounting system reviews and pre-award audits, and Cost Accounting Standards (CAS) Statement reviews. We selected these elements because they were identified by the NSF-OIG as deficiencies in the past and are essential to determining that the contractor has the ability to complete the contract cost requirements.
- Contract monitoring. We focused on the development of monitoring or surveillance plans, monitoring activities, and incurred cost audits. These activities were previously identified by the NSF-OIG as deficiencies and are key to determining if the contractor is performing as expected and within allowable costs.

To determine key practices in each of these areas, we relied on prior reports and findings from the GAO, NSF-OIG, and other agencies. Below is the list of GAO reports we relied on:

- GAO, *Standards for Internal Control in the Federal Government*, GAO/AIMD-00-21.3.1 (Washington, D.C.: November 1999);
- GAO, *Contract Management: Trends and Challenges in Acquiring Services*, GAO-01-753T (Washington, D.C. May 22, 2001);
- GAO, *Defense Contracting: Improved Insight and Controls needed Over DOD's Time-and-Materials Contracts*, GAO-07-273 (Washington, D.C.: June 29, 2007);
- GAO, *Contract Management: Extent of Federal Spending under Cost-Reimbursement Contracts Unclear and Key Controls Not Always Used*, GAO-09-921(Washington, D.C.: Sept. 30, 2009); and
- GAO, *Acquisition Planning: Opportunities to Build Strong Foundations for Better Services Contracts*, GAO-11-672 (Washington, D.C.: Aug. 9, 2011).

We also reviewed internal NSF guidance and the Federal Acquisition Regulation (FAR) for additional key practices.

To determine the extent to which NSF's contracting practices incorporate key practices and address prior NSF-OIG recommendations, we reviewed a nongeneralizable sample of 11 contracts and orders with funding obligations over $3 million in fiscal year 2011, the latest year for which data were available when we began our work.[14] We used a risk-based approach to select our sample to ensure it included NSF acquisitions with the highest obligation dollar amount. The 11 contracts and orders selected for review represent 70 percent of total contract obligations in fiscal year 2011 and reflect a mix of program offices, a range of obligation amounts, and a variety of contract types, such as fixed-price and cost-reimbursement. We selected four contracts for which NSF used the negotiation process set forth in Part 15 of the Federal Acquisition Regulation and seven orders on existing contracts for which NSF used streamlined procedures described in other parts of the FAR. The four negotiated acquisitions in our sample are cost-reimbursement contracts and represent about 56 percent of NSF's total fiscal year 2011 contract obligations and about 80 percent of the obligations in our sample. The seven streamlined acquisitions represent about 14 percent of NSF's fiscal year 2011 contract obligations and 20 percent of the obligations in our sample. One of the seven streamlined orders is a hybrid contract type using fixed-price and time-and-materials (T&M) elements; one is a cost- reimbursable order; and the other five are T&M orders. Although the 11 contracts were active during the time of our review, some of the selected contracts were awarded more than 7 years ago—before NSF updated its contracting manual to provide more procedural guidance— and some more recently.

We reviewed the files for the selected contracts and used practices identified in the FAR, NSF internal guidance, and prior GAO reports to assess NSF's use of key practices and procedures for the acquisition planning, award, and contract monitoring phases. In addition to contract file review, we met with contract and program officials to confirm our understanding of information in the contract files and of NSF's practices and procedures as evidenced by the contract files. We also reviewed and considered additional documentations provided by the program and contract officials that were not maintained in the contract files.

To assess progress NSF made in response to prior NSF-OIG findings, we reviewed prior NSF-OIG recommendations and corrective action plans. We met with NSF-OIG officials to better understand their recommendations related to our review and used this information to provide assessments of progress made in response to these findings.

NSF was in the process of a full acquisition system assessment when we initiated our review. While we were completing our audit work, NSF issued a review of its acquisition function in July 2012. While we met with the internal controls officials involved in this review to understand their process, we did not assess the NSF review as part of this review.

We conducted this performance audit from February 2012 to March 2013 in accordance with generally accepted government auditing standards. Those standards require that we plan and perform the audit to obtain sufficient, appropriate evidence to provide a reasonable basis for our findings and conclusions based on our audit objectives. We believe that the evidence obtained provides a reasonable basis for our findings and conclusions based on our audit objectives.

End Notes

[1] One of the contracts we selected based on 2011 obligations had expired and was replaced with a new contract awarded in December 2011. To provide a more current assessment we reviewed the contract awarded in fiscal year 2012.

[2] GAO, *Acquisition Planning: Opportunities to Build Strong Foundations for Better Services Contracts*, GAO-11-672 (Washington, D.C.: Aug. 9, 2011); and FAR 7.102.

[3] The Federal Supply Schedules program consists of contracts awarded by GSA or the Department of Veterans Affairs for similar or comparable goods or services, established with more than one supplier, at varying prices. Federal Acquisition Regulation (FAR) § 8.401 and § 8.402. The program offers a large group of commercial products and services ranging from office supplies to information technology services.

[4] GAO, *Contract Management: Extent of Federal Spending under Cost-Reimbursement Contracts Unclear and Key Controls Not Always Used*, GAO-09-921 (Washington, D.C.: Sept. 30, 2009).

[5] GAO-09-921.

[6] Prior to the October 2012 Acquisition Manual, the "requirements package" was known as the "request for contract" package.

[7] While assessing risk and identifying opportunities to use a less risky contract is a sound practice, there was no requirement to document this analysis prior to FAR changes effective March 16, 2011. 76 Fed. Reg. 14,543 (interim rule). 77 Fed. Reg. 12,925 (March 2, 2012) (final rule, effective April 2, 2012)). FAR § 16.103(d).

[8] GAO-09-921; GAO, *Defense Contracting: Improved Insight and Controls needed Over DOD's Time-and-Materials Contracts*, GAO-07-273 (Washington, D.C.: June 29, 2007); DOD, September 2004 Defense Procurement and Acquisition Policy Memo, *Requirements for Service Contracts;* and FAR Part 16.

[9] GAO-09-921.

[10] Cost Accounting Standards generally apply to cost-reimbursable contracts above a certain dollar threshold.

[11] GAO-09-921; GAO, *DCAA Audits: Widespread Problems with Audit Quality Require Significant Reform*, GAO-09-468 (Washington, D.C.: Sept. 23, 2009); *Defense Contracting: DOD Initiative to Address Audit Backlog Shows Promise, but Additional Management Attention Needed to Close Aging Contracts*, GAO-13-131 (Washington, D.C.: Dec. 18, 2012).

[12] Incurred cost audits are an important tool that enables management to assess a contractor's compliance with the financial terms and conditions of a contract.

[13] GAO-13-131.

[14] One of the contracts we selected based on 2011 obligations had expired and was replaced with a new contract awarded in December 2011. To provide a more current assessment we reviewed the contract awarded in fiscal year 2012.

INDEX

D

E

F

G

H

I

O

P

Q

R

S

T